Christianson Educators

SAT® MATH

Made Visual

To learn more, visit www.mathmadevisual.com.

Paperback: ISBN 979-8-9931178-0-5
Ebook: ISBN 979-8-9931178-2-9

Library of Congress Control Number: 2025920097

First Edition, 2025

Design and illustrations by Noel Anderson.
Edited by Karyl Garland.
Mathematical review by S. Janani Lakshmanan, PhD in Mathematics.

This book's full title is *Christianson Educators SAT Math Made Visual;* throughout the text it is referred to simply as *SAT Math Made Visual.*

Contents

Hello there! I'm Desmoto. I'm a coordinate point with a sharp eye for graphs and a soft spot for bowties. I'll be your guide, sharing tips to help you get the most out of Desmos.

Introduction

Math has a strong visual component. It appears in patterns, shapes, and relationships—and a graph is one of the clearest ways to represent those relationships. A graph helps you see how values change, compare, or intersect, making the behavior of an equation easier to understand.

This visual perspective is supported by Desmos, a free online calculator. It generates graphs instantly and handles the algebra behind them, giving you more room to focus on what the graph reveals. On the SAT exam, Desmos can simplify many problems and help you find answers more efficiently.

Now that the SAT exam has gone digital, every student has access to the Desmos graphing calculator. This guide shows you how to use it effectively. Inside are six core techniques and 50 targeted exercises that highlight a visual approach that can reveal answers quickly. Whether you choose to solve problems algebraically, visually, or both, this book is designed to give you options and support your best work on test day.

Enter information here.

www.Desmos.com

Graphs appear here.

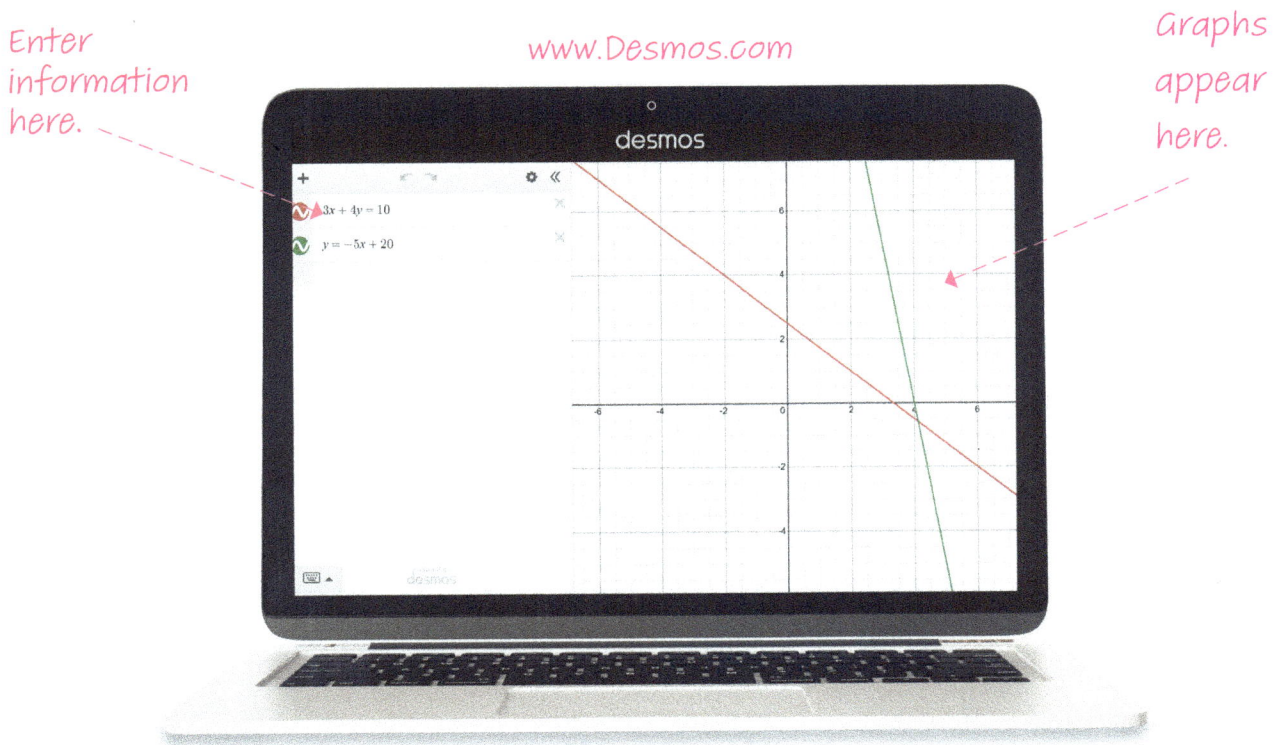

Desmos is a free, web-based calculator. It is also built into the digital SAT exam.

Your Desmos Toolkit

Your Desmos Toolkit

Here are the core Desmos tools you'll use in the book. Don't worry if everything doesn't make sense at first—some ideas, like sliders, are best understood by using them in the exercises. This section introduces the tools, but real understanding builds as you practice.

Accessing the Desmos Calculator

Option	Link/Access
SAT Version	https://www.desmos.com/testing/cb-digital-sat/graphing
Standard Version	Go to https://www.desmos.com and select **Graphing**.
Desmos App	Available on the *iOS App Store* and *Google Play Store*

Note that the SAT version of the Desmos calculator differs slightly from the standard one. For this guide, either version works.

The Expression List

The exercises in *SAT Math Made Visual* are based on two core steps: entering information into the Expression List and interpreting the graphs that follow. Getting comfortable with the Expression List matters because it's where each question begins.

Entering Information

When you enter an equation in the Expression List, its graph appears in the Graph Window and matches the color of the **Color Icon** 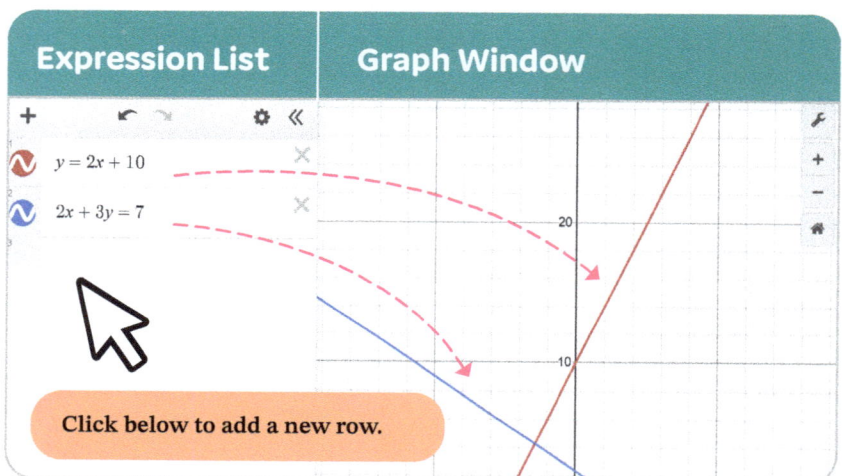. Each equation should be entered on its own row.

Three Ways to Add a Row

1. Click below an existing row.
2. Click an existing row, then press **Return** or **Enter.**
3. Click **Add Item** ⊕ and select $f(x)$ **expression** from the dropdown menu.

Expression List **Graph Window**

$y = 2x + 10$

$2x + 3y = 7$

20

10

Click below to add a new row.

Tips for Entering Equations

Using Standard Variables
Desmos may recognize other letters as variables, but it's safest to use the standard variables x and y in lowercase form.

Typing
Check what you type! Mistyping information is a common mistake.

Warning Signs
If a **Warning Sign** ⚠ appears instead of a **Color Icon** 🅝, there's a problem with what was entered and the graph will not appear. Click ⚠ to find out what's wrong and how to fix it.

Inputting Large Numbers
Avoid using commas in large numbers. For example, write 50000 instead of 50,000.

The Desmos Workspace

Here are some key features of the Expression List and Graph Window.

Key	Expression List	Graph Window

Edit List ⚙ opens options for the selected equation, including creating a table of values. This guide refers to this automatically generated table as a Function Table.

Add Item ➕ adds a new row or a regular table for typing in your own coordinate points.

Color Icon 🌀 opens options like changing graph colors.

Undo/Redo ↶ ↷ reverses or restores actions.

Hide/Show Expression List « » shows or hides the Expression List.

Delete Row ✕ removes a row.

Keypad ⌨ opens the Desmos Keypad.

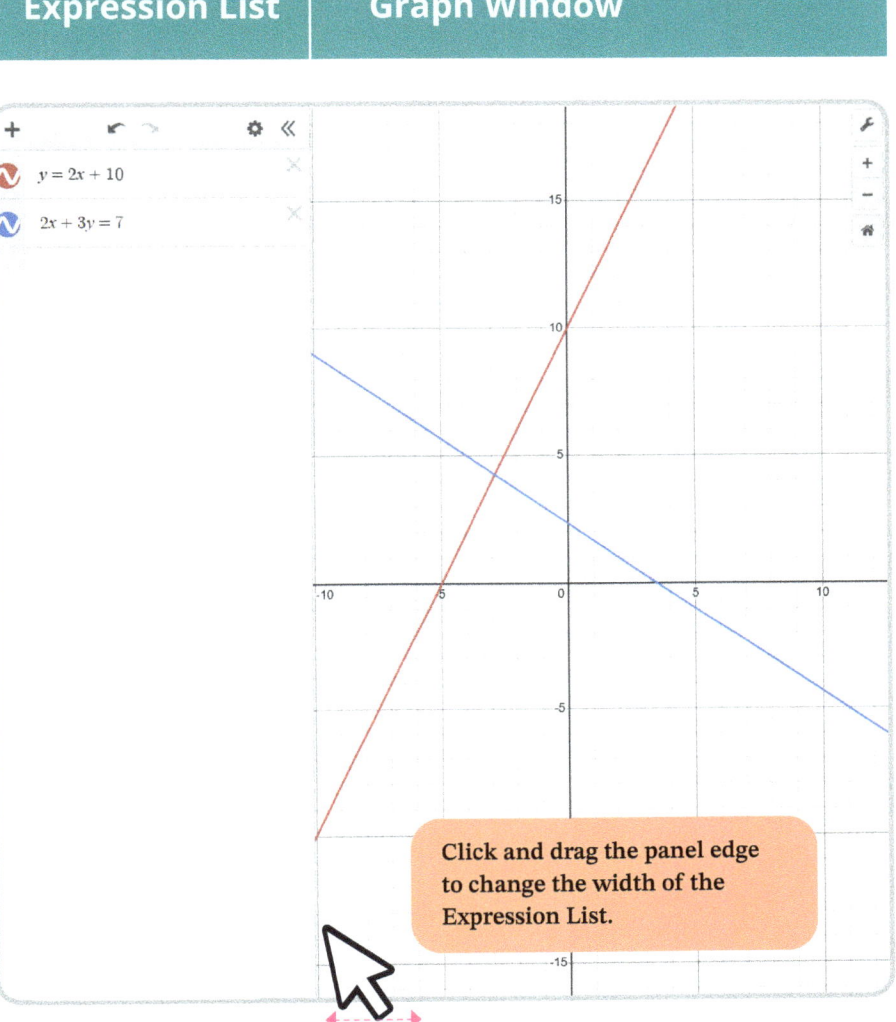

$y = 2x + 10$

$2x + 3y = 7$

Click and drag the panel edge to change the width of the Expression List.

Graph Settings 🔧 opens graph settings.

Zoom In ➕ /**Out** ➖ changes the view of the Graph Window.

Default View 🏠 resets the graph to its original position. If the house icon is absent, the graph is already in its default view.

Plotting Coordinate Points

Sometimes you'll need to enter coordinate points (also called points or coordinates) into the Expression List. Here are two basic methods.

Method 1

Enter coordinate points using a table.

 **Step 1** Click **Add Item** ⊕ and then select **Table** 📖.

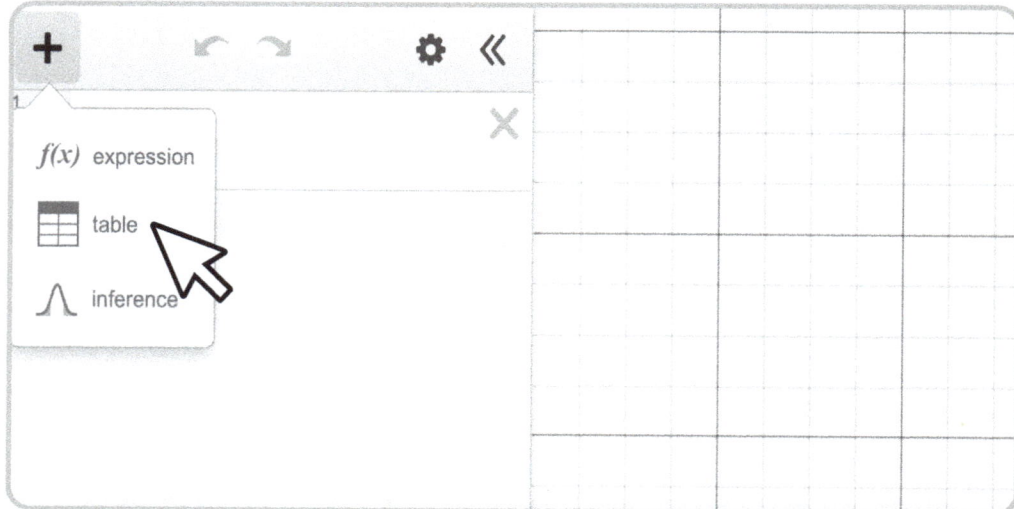

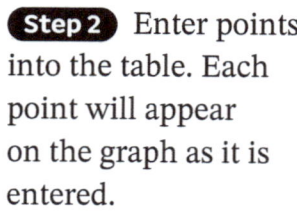 **Step 2** Enter points into the table. Each point will appear on the graph as it is entered.

Clicking 🔍 adjusts the window to fit all coordinate points.

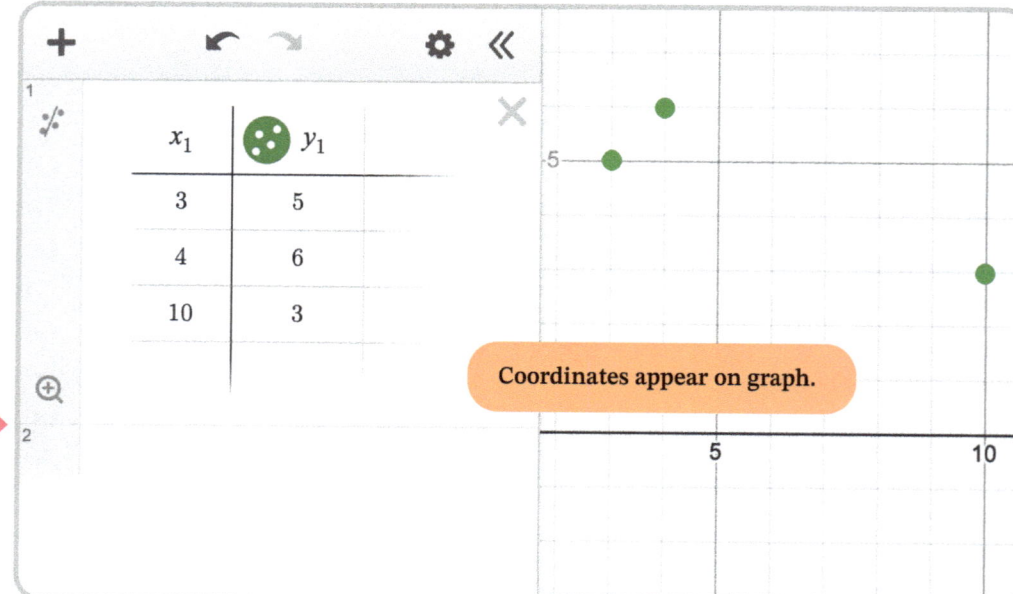

Coordinates appear on graph.

Method 2

Enter coordinate points into the Expression List.

Enter coordinate points directly into the Expression List, separating each point with a comma. To label points, check the box next to **Label**.

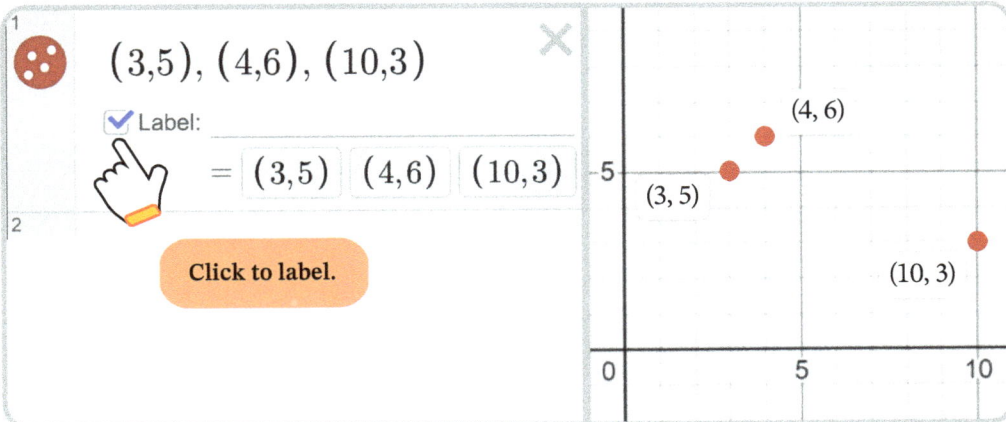

Connecting Coordinate Points

When points are entered on the same row, you can connect them. Click and hold the **Color Icon** 〰 (icons may look different, but they work the same way). Open the Color Panel, then click the **Lines** button to connect the points.

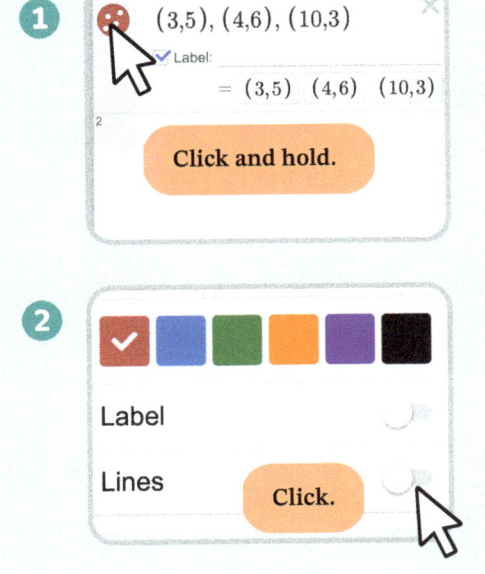

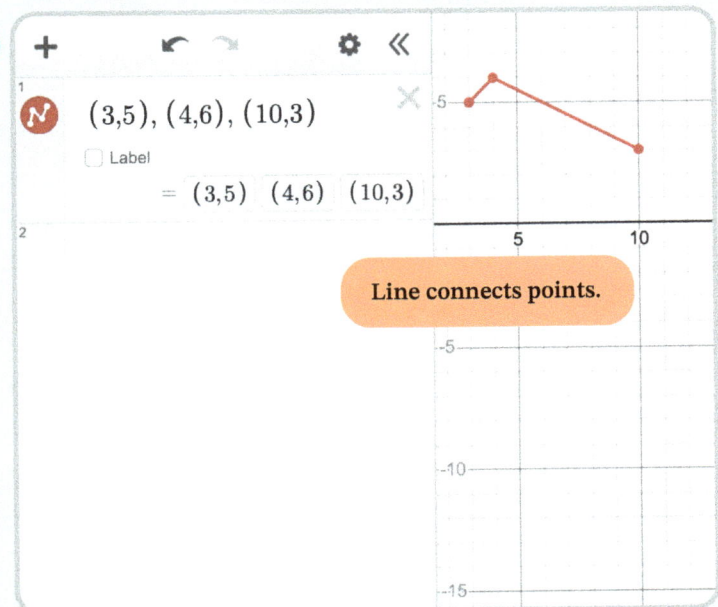

Sliders

Sliders let you assign and adjust values for unknown constants so you can explore how different values affect an equation.

An equation with an unknown constant, like $3x + 4c = 5y$, cannot be graphed until that constant has a value. That's where sliders come in. A slider lets you set and adjust the value for c (or another constant), and Desmos instantly updates the graph to reflect the change.

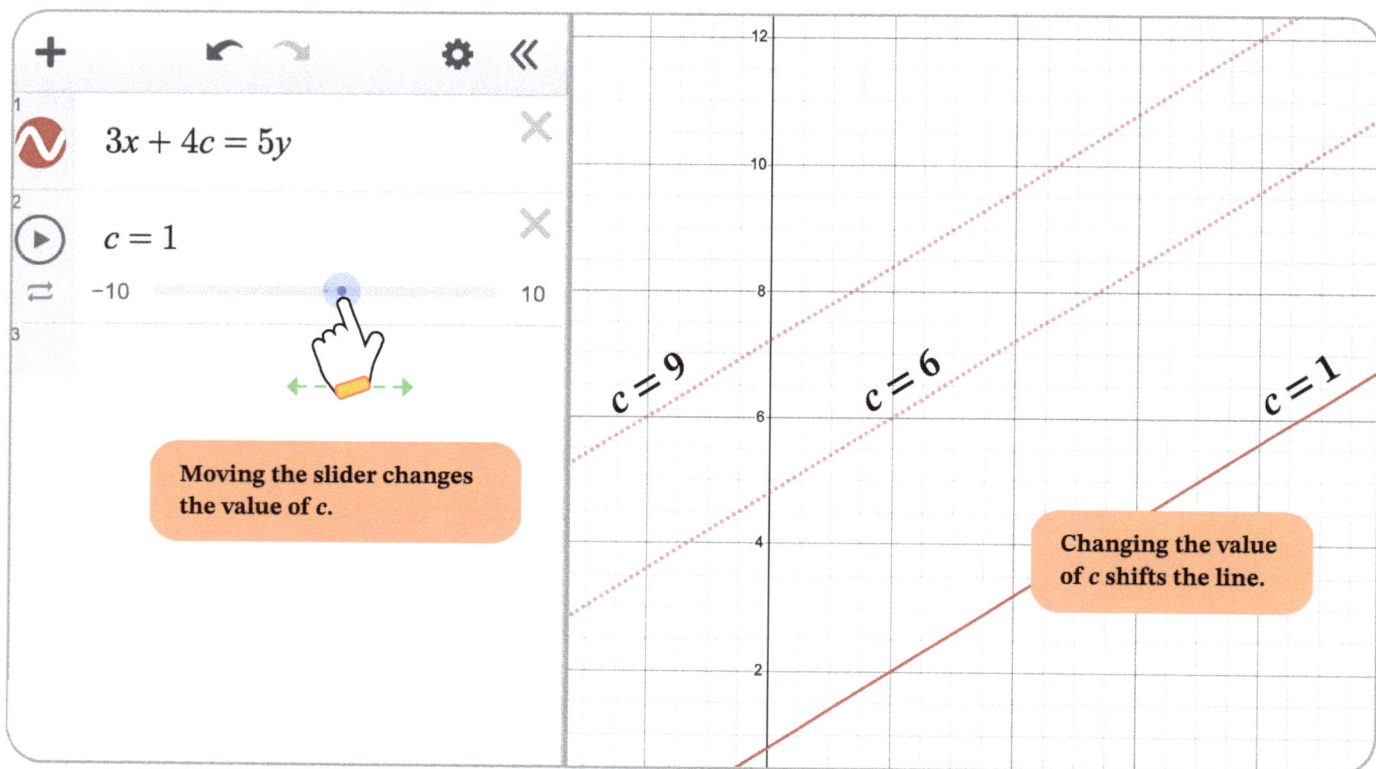

Moving the slider changes the value of c.

Changing the value of c shifts the line.

How to Add a Slider

Accepting a Slider

Desmos will offer a slider for an unknown constant. Accept it by clicking the blue button c.

Adjusting a Slider

Drag the blue dot or delete the slider's current value and type in a new one. The graph will update as the value of the constant changes.

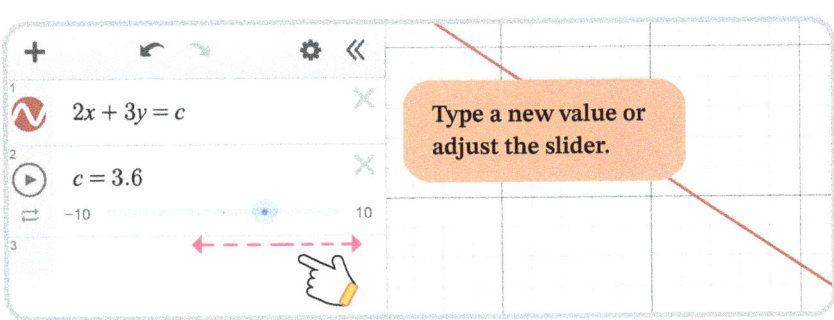

Changing the Range

To adjust the default range, $-10 \leq c \leq 10$, click anywhere in the slider's row and enter a new range, placing the lower value to the left of c. Click outside the row to apply the change.

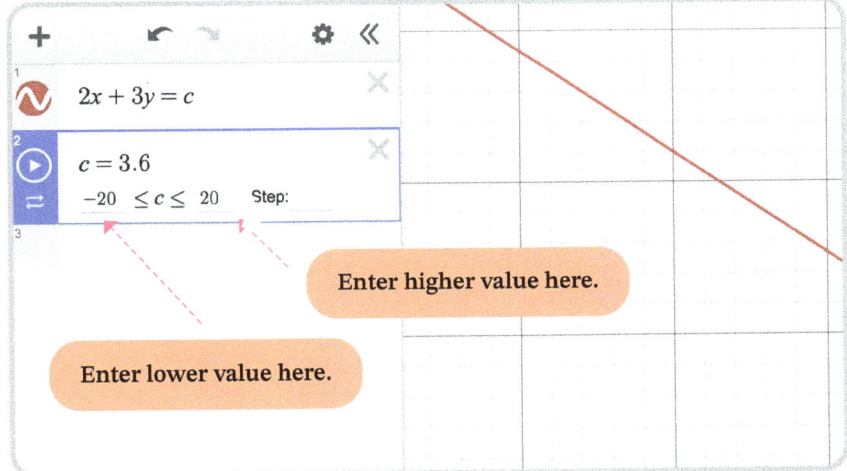

Finding the Right Range

Finding the right slider range may take a few tries. Use a wider range when you're unsure where to start, then narrow it to fine-tune decimal values. For example, to reach 24.75, you might set the range to $24 \leq c \leq 25$.

Sliders

Where's the Slider?

Desmos sometimes doesn't recognize that you need a slider and instead treats the unknown value as a variable. For example, if you type $y = a$, Desmos treats a as a variable and graphs $y = x$. That's not what you want.

The example below shows what this issue looks like and how to fix it.

Prompting Desmos to Add a Slider

Without a slider, $y = a$ is graphed incorrectly as $y = x$.

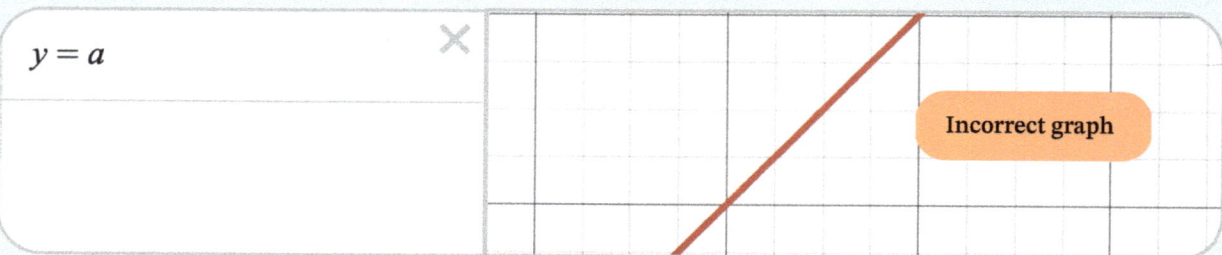

To fix this situation, type $a = 4$ (or any value) in a new row. Desmos will automatically create a slider for a and graph the equation correctly.

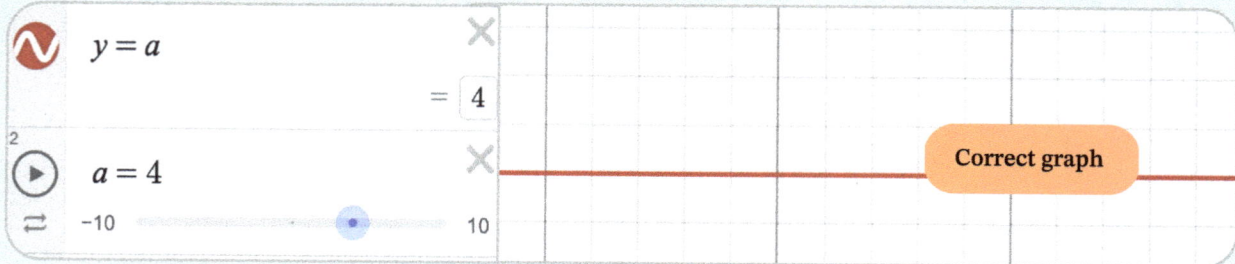

Slider Troubleshooting

Sliders can be confusing at first. Later exercises will walk you through them. This page is here as a reference if something doesn't behave the way you expect.

Why does Desmos automatically provide a slider in certain situations?

Desmos adds sliders for equations like $y = 0$. If you don't need the slider, just ignore it.

What if the slider option disappears?

If the slider prompt vanishes, delete part of the equation until it reappears or follow the directions on page 10.

How do sliders handle repeating fractions?

Sliders can't stop exactly on repeating decimals. To test a value like $\frac{2}{3}$, type the fraction directly (for example, $c = \frac{2}{3}$).

How do I restore the slider?

If the slider disappears after entering a fraction, type a whole number or decimal to restore it.

Do I always have to accept a slider?

No. Sometimes Desmos offers a slider when you don't need one. For example, if you're typing a function like $f(x)$, Desmos may ask to create a slider for f. Ignore the prompt. Once you finish typing the full equation, Desmos will recognize it as a function and the slider option will disappear.

Typing Tips

You'll be entering a lot of information into the Expression List. Even small typing errors can change the graph or give the wrong answer, so accuracy matters. Here are some tips for entering data and symbols correctly.

Staying In or Exiting an Operation

In Desmos, staying in or exiting an operation depends on how your expression is typed, and that choice affects the final results.

To Stay in an Operation

Use parentheses to keep everything grouped correctly.
Example: Enter 3^{x+5} as $3^{(x+5)}$

To Exit an Operation

Use the right arrow key $\boxed{\rightarrow}$
to move out of an operation.
Example: Enter $\sqrt{4} + 6$ as $\sqrt{4}$ $\boxed{\rightarrow}$ $+ 6$

> Made a mistake while typing an equation? Use your mouse or the left and right arrow keys $\boxed{\leftarrow}$ $\boxed{\rightarrow}$ to move to the spot you want. Then erase and retype from there.

Desmos Keyboard

Click ⌨ to open the Desmos keyboard.

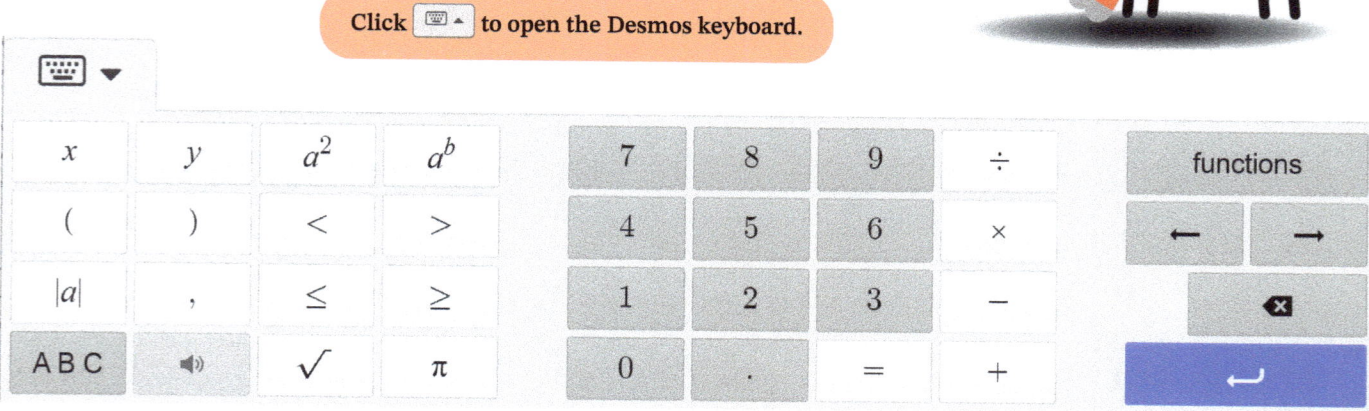

Typing Common Expressions

Math Expressions	Desmos Keypad	Computer or Tablet Keyboard
Fractions e.g., $\dfrac{x+4}{5} + 6$	» Click $(~$, $~x~$, $~+~$, $~4~$, $~)$ » Click $\div$ » Click 5 » Click $\longrightarrow$ » Click $+$, 6	» Type $(x+4)$ » Type / (forward slash key) » Type 5 » Type $\rightarrow$ » Type + 6
Absolute Value e.g., $\|x-4\|$	» Click $\|a\|$ » Click x, $-$, 4 » Click $\|a\|$	» Type | (pipe key) » Type $x - 4$ » Type |
Squares e.g., x^2	» Click x » Click a^2	» Type x » Type ^ (caret key) » Type 2
Exponents e.g., x^4	» Click x » Click a^b » Click 4	» Type x » Type ^ (caret key) » Type 4
Square Roots e.g., $\sqrt{5} + 2$	» Click $\sqrt{~}$ » Click 5 » Click $\longrightarrow$ » Click $+$, 2	» Type *sqrt* » Type 5 » Type $\rightarrow$ » Type + 2
Nth Roots e.g., $\sqrt[3]{8}$	» Click functions » Click $\sqrt[n]{~}$ » Click 3 » Click $\longrightarrow$ » Click 8	» Type *nthroot* » Type 3 » Type $\rightarrow$ » Type 8

Navigating Graphs

Graphs are central to visual problem solving in Desmos, so knowing how to navigate them is essential. On the real SAT exam, the calculator doesn't open full screen, so being able to move around the graph efficiently is especially important.

Identifying Coordinate Points

Once you enter an equation in the Expression List, its graph will appear. For most graphs, you can click directly on a graph to view its coordinates or click, hold, and drag along the graph to see the coordinates update automatically. On tablets, tap the figure first, then click the coordinate point.

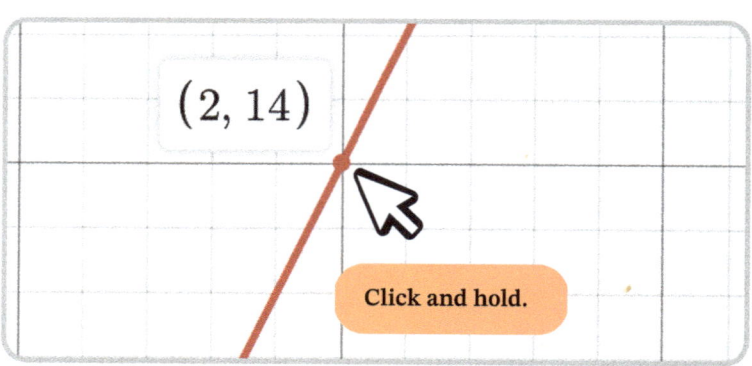

Hiding Graphs

Click the **Color Icon** 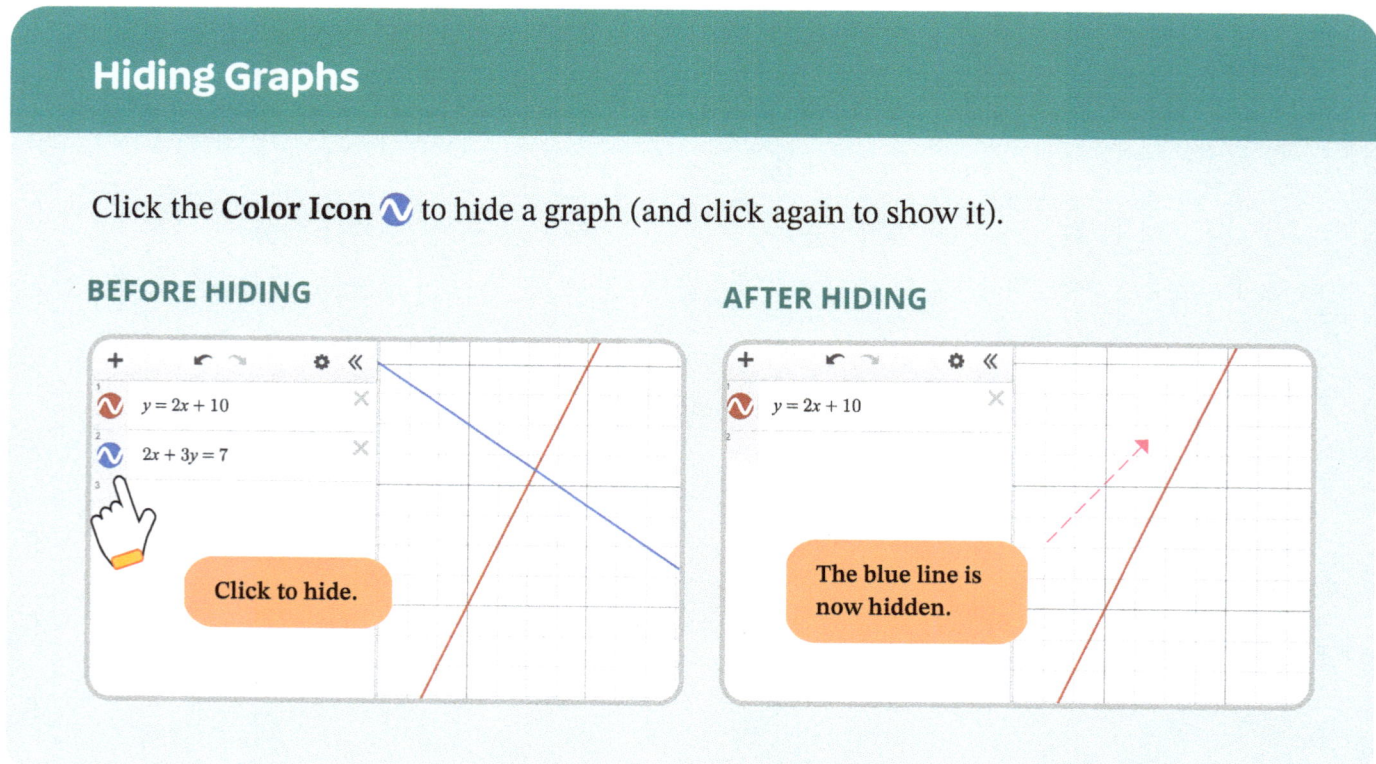 to hide a graph (and click again to show it).

BEFORE HIDING | AFTER HIDING

The blue line is now hidden.

The Graph Window

The Graph Window is where you'll view and explore graphs in Desmos.

Points of Interest

Points of interest are key coordinates on a graph, such as intercepts, intersections, and maximum or minimum values.

To view them, click directly on that graph. Desmos will display the points (in gray or black) for that graph only.

Moving the Window

Click and drag the Graph Window to move it. On tablets, drag your finger across the screen.

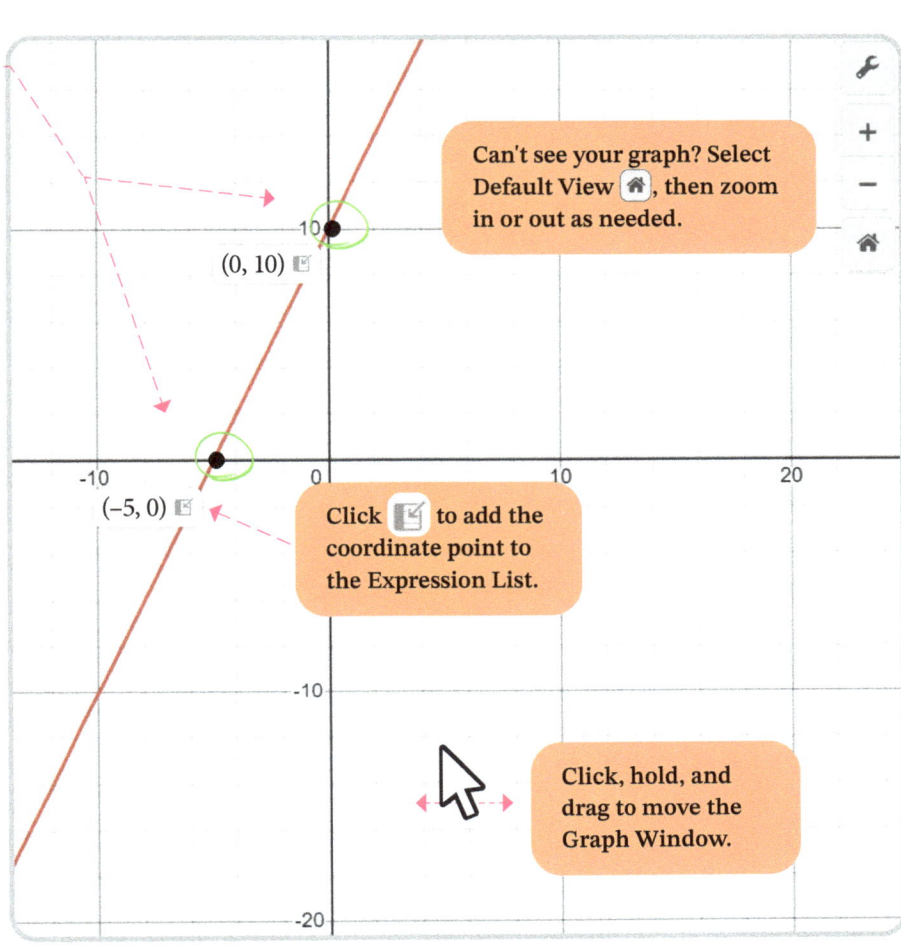

Can't see your graph? Select Default View 🏠, then zoom in or out as needed.

(0, 10)

(−5, 0)

Click ☑ to add the coordinate point to the Expression List.

Click, hold, and drag to move the Graph Window.

Resetting the Window

Click **Default View** 🏠 to reset the Graph Window. If the house icon is missing, you're already in the default view.

Zoom In ⊕ /Out ⊖

Use these tools to navigate graphs.

To get the clearest view, make the Desmos interface larger by dragging the corner of the screen.

Key References

The references below are for this book. On test day, remember to check the references you're given.

Graphing References

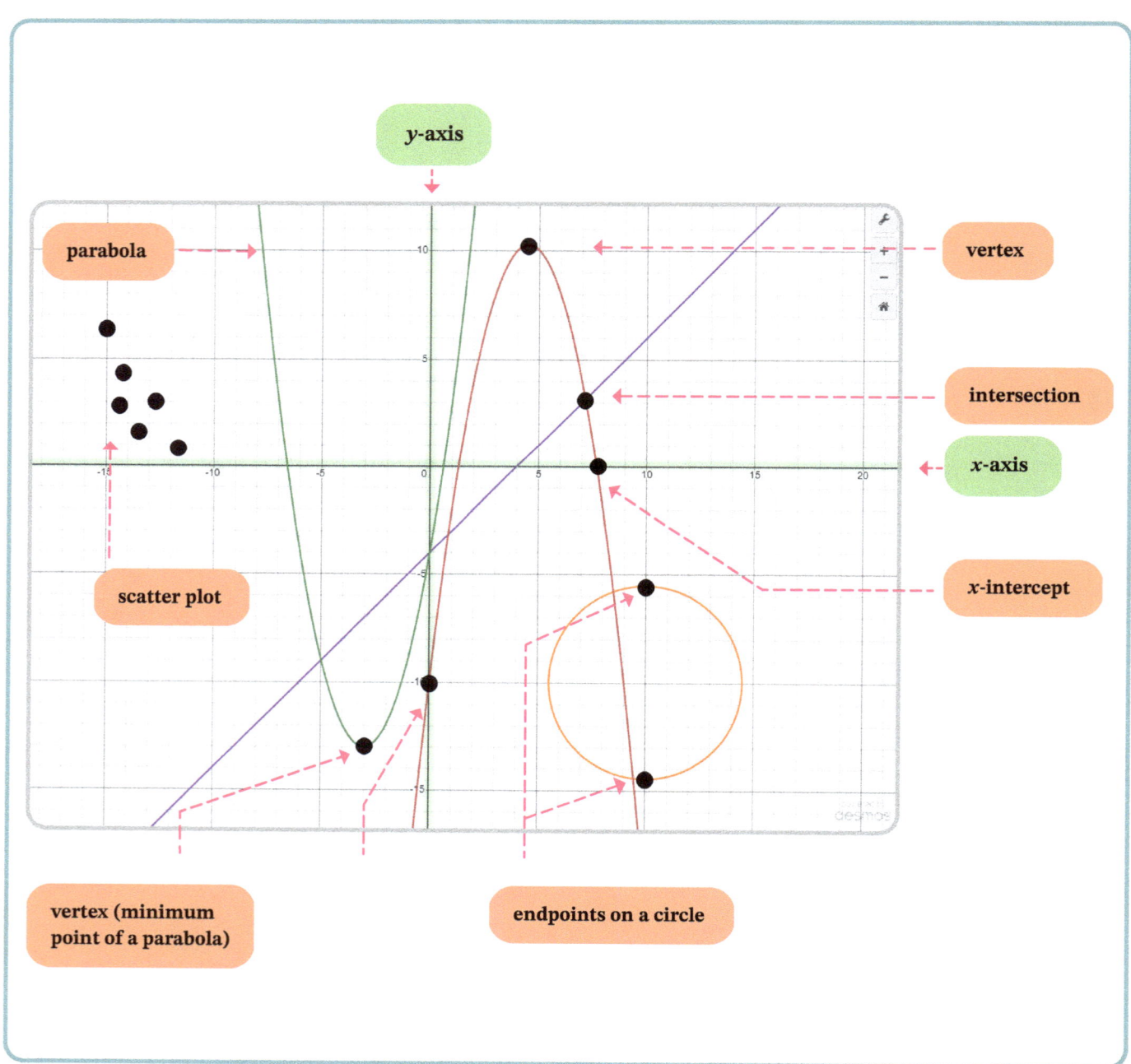

Equation Forms

Equation	Form	Constants and Coefficients
$y = mx + b$	Slope-Intercept Form/Line	m is the slope b is the y-intercept
$y = ax^2 + bx + c$	Standard Form/Quadratic	a is the coefficient of x^2 ($a \neq 0$) b is the coefficient of x c is a constant
$y = a(x - h)^2 + k$	Vertex Form/Quadratic	a controls the width/direction (h, k) is the vertex
$(x - h)^2 + (y - k)^2 = r^2$	Standard Form/Circle	(h, k) is the center r is the radius

Geometry References

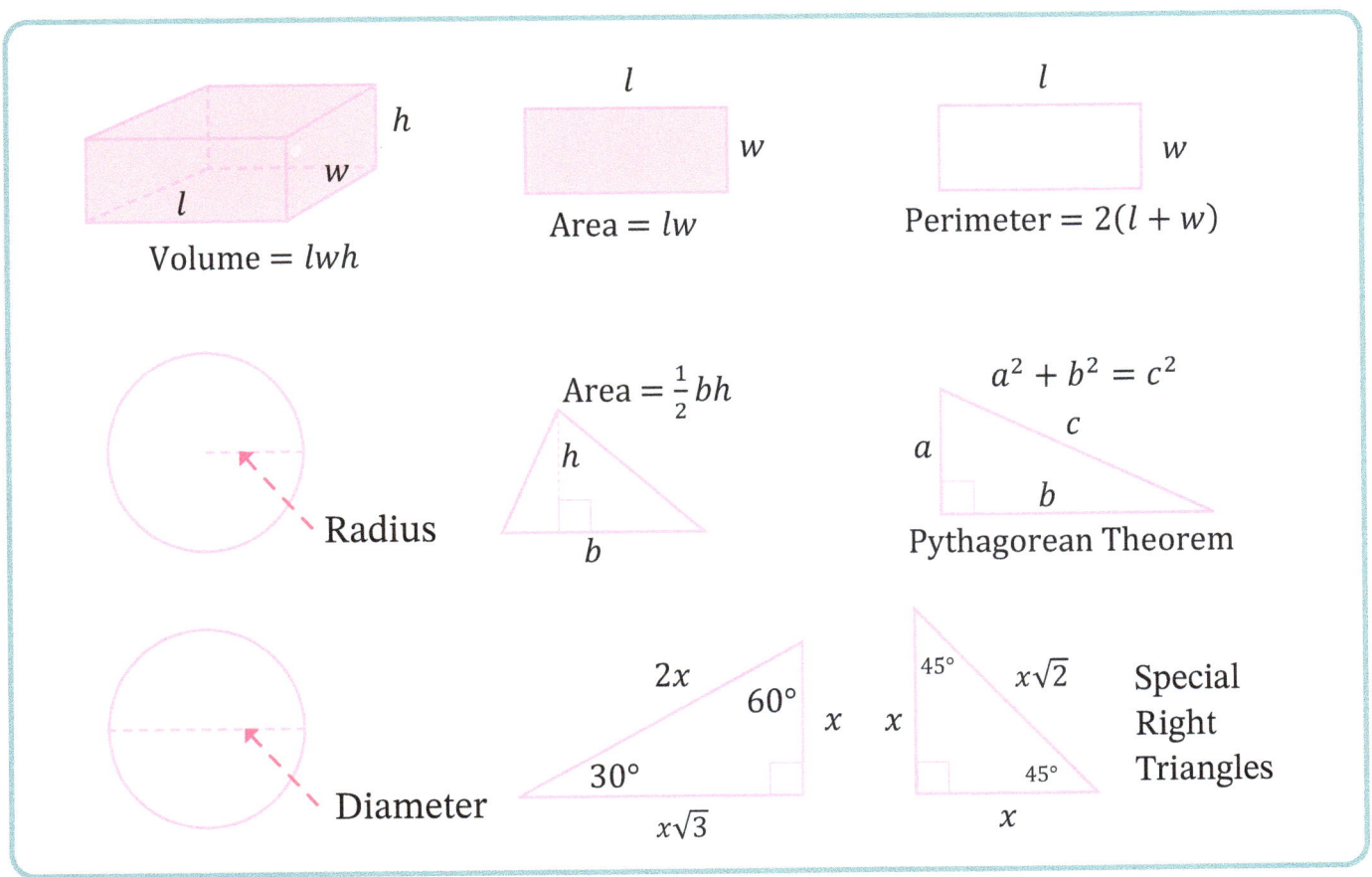

Basic Calculations

This page introduces a few other calculator tools. Some appear in the exercises, while others are included simply to help you recognize what Desmos can do.

Desmos as a Calculator

Math Operations

Desmos can perform basic arithmetic.

$3 \cdot 89$

$= 267$

Percents

Desmos automatically adds the word *of* when you enter the % sign.

25% of 41

$= 10.25$

Decimals/Fractions

Select  to turn a decimal into a fraction or vice versa.

.45

$= \frac{9}{20}$

Stats and Geometry

Enter data or coordinates as shown. Type **mean, median, distance,** or **midpoint,** or choose the function from the **Functions** menu on the Desmos Keypad. The answer appears in a gray box or on the graph.

mean $(3,6,5,10)$

$= 6$

median $(25,3,5,11,24)$

$= 11$

distance $((2,6),(5,10))$

$= 5$

midpoint $((2,4),(6,50))$

☐ Label $= (4,27)$

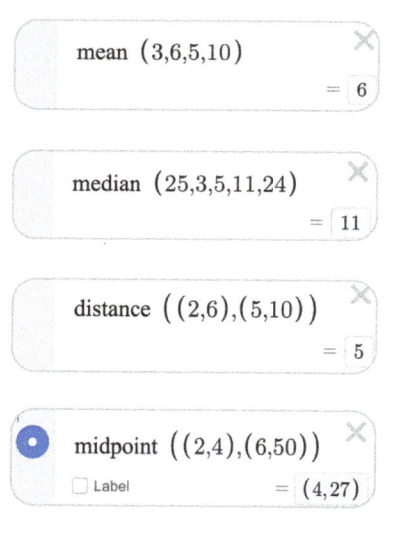

Bringing a handheld calculator to the SAT exam gives you options. You can choose whichever tool feels faster or clearer for a particular question.

Navigating the Exercises

We're just about ready to dive into the exercises. Each one follows a consistent format, shown below. As you work through the exercises, enter each step into Desmos rather than just reading along. Doing every exercise builds fluency. Complete them even if you already know the algebra.

Features	Layout

Tips and notes help you avoid common mistakes.

Question and answer choices mimic real questions.

Highlights show connections between the question and calculator.

Steps show how to solve.

Green circles lead to final answer.

Desmoto shares extra tips.

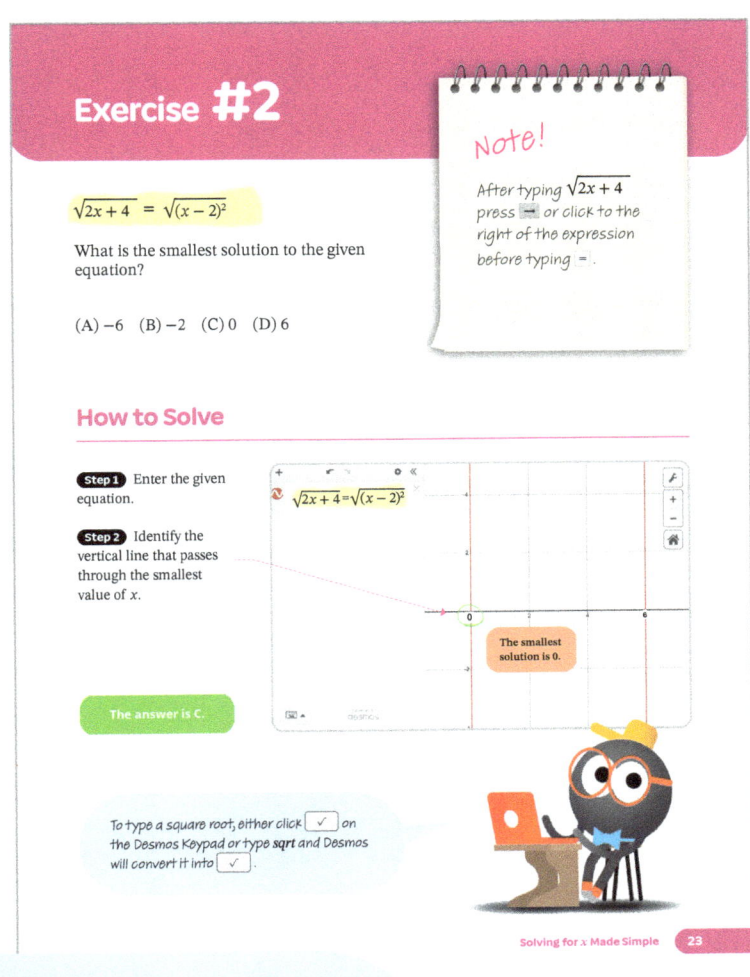

Now turn the page to learn about Technique #1. Got your thinking cap on? I do.

Technique

#

Solving for x Made Simple

Solving for x Made Simple

Welcome to Technique #1, where you'll learn how to solve for x by reading vertical lines on a graph. The overall approach stays the same across Exercises 1–8, but each problem gives you practice entering different expressions and using Desmos tools. Working through each one builds fluency with both the technique and the calculator.

Example

Let's walk through the steps using a simple example: solving for x in the equation $2x - 4 = 10.$

Step 1 Enter $2x - 4 = 10$ into a row.

Step 2 Identify the solution by reading the equation of the vertical line. All points on the line have an x value of 7, so the vertical line is $x = 7$. Therefore, the solution to $2x - 4 = 10$ is $x = 7$.

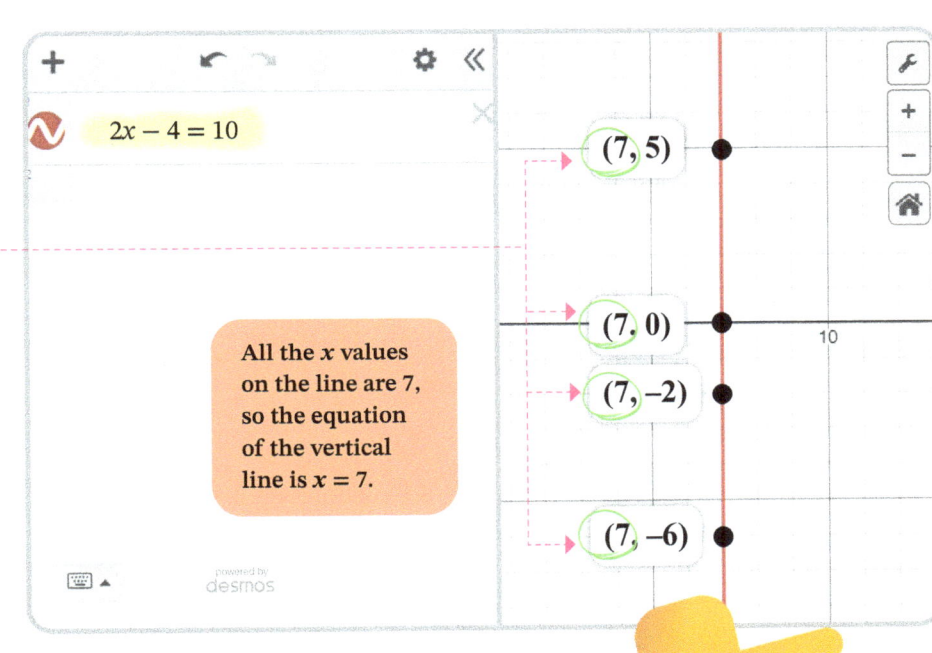

All the x values on the line are 7, so the equation of the vertical line is $x = 7$.

Who says simple can't be brilliant? Watch what happens when you let the graph do the work.

Exercise #1

$$x^3 + x^2 - 4x - 4 = 0$$

What is one solution to the equation above?

_____

Note!

To access Desmos, go to **www.desmos.com** and select **Graphing**. Enter each step as you follow along. For other access options, see page 3.

How to Solve

Enter information in Desmos.

Graphs appear here.

Step 1 Enter the given equation.

Step 2 Identify the x values of the vertical lines: -2, -1, and 2. Choose one of these x values as your answer.

$x^3 + x^2 - 4x - 4 = 0$

zoom in!

−2   −1   0   1   2

Choose −2, −1, or 2.

The answer is −2, −1, or 2.

powered by desmos

Clear the row by clicking **Delete Row**  before moving to Exercise #2.

Exercise #2

$$\sqrt{2x + 4} = \sqrt{(x - 2)^2}$$

What is the smallest solution to the given equation?

(A) −6 (B) −2 (C) 0 (D) 6

Note!

After typing $\sqrt{2x + 4}$ press → or click to the right of the expression before typing =.

How to Solve

Step 1 Enter the given equation.

Step 2 Identify the vertical line that passes through the smallest value of x.

$$\sqrt{2x + 4} = \sqrt{(x - 2)^2}$$

The smallest solution is 0.

The answer is C.

powered by desmos

To type a square root, either click ✓ on the Desmos Keypad or type **sqrt** and Desmos will convert it into ✓ .

Exercise #3

$2|3x + 3| = 6$

What is the value of x if $x < 0$?

____

Note!

To enter an absolute value sign, click |a| on the Desmos Keypad.

How to Solve

Step 1 Enter the given equation.

Step 2 Find the vertical line through a negative value of x. The equation of the line is $x = -2$.

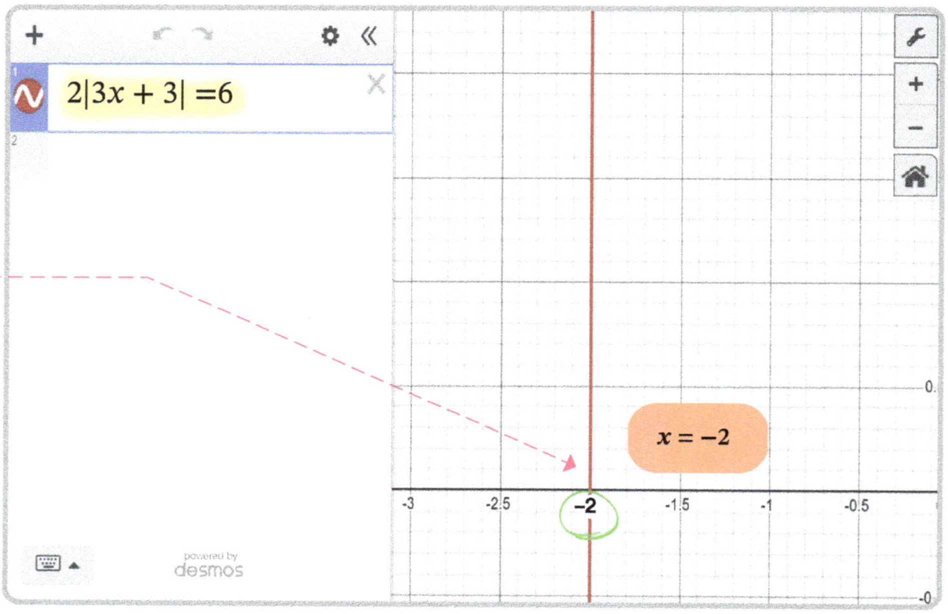

2|3x + 3| =6
x = −2
-3
-2.5
−2
-1.5
-1
-0.5
powered by desmos

The answer is -2.

Utilize the **Zoom In** *and* **Zoom Out** *tools to see points or graphs more clearly.*

Exercise #4

23% of a number is 115. What is the number?

(A) 61 (B) 500 (C) 115 (D) 200

Note!

Write the equation as follows:

23% of x = 115

How to Solve

Step 1 Enter the equation as shown.

Step 2 Click the x-intercept to find the equation of the vertical line, $x = 500$.

The answer is B.

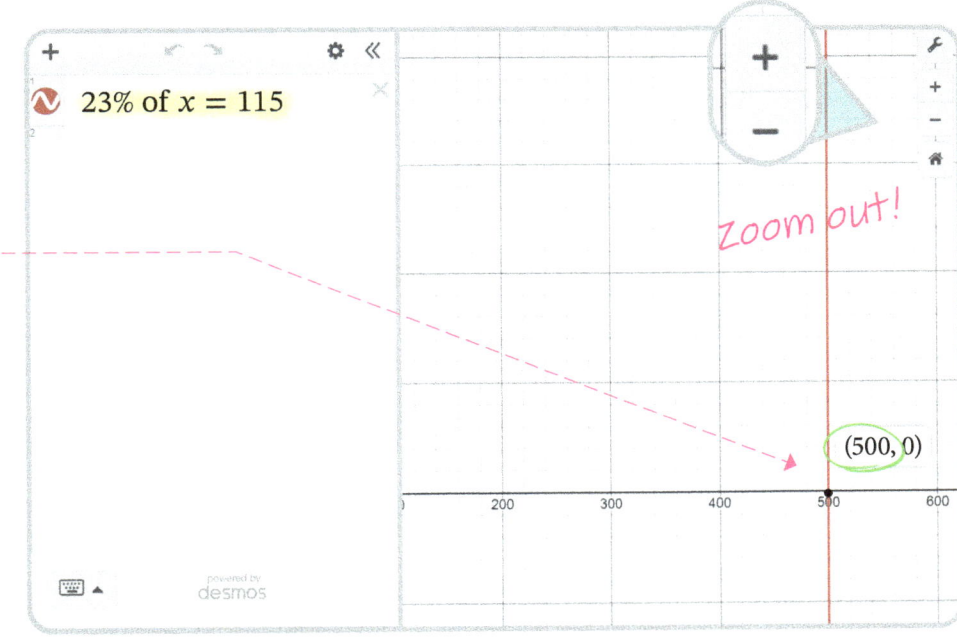

23% of x = 115

Zoom out!

(500, 0)

200   300   400   500   600

desmos

Click **Default View** 🏠 at the end to reset the graph to its original position.

$$5, 6, x + 1, \frac{3x + 15}{4}$$

The average (arithmetic mean) of the data above is x. What is the value of x?

(A) 4   (B) 5   (C) 6   (D) 7

Note!

Enter the following equation, including the word **mean**:

$$\text{mean}\left(5, 6, x + 1, \frac{(3x + 15)}{4}\right) = x$$

How to Solve

Step 1 Enter the equation as shown.

Step 2 Identify the x-intercept of the vertical line. The equation of the line is $x = 7$.

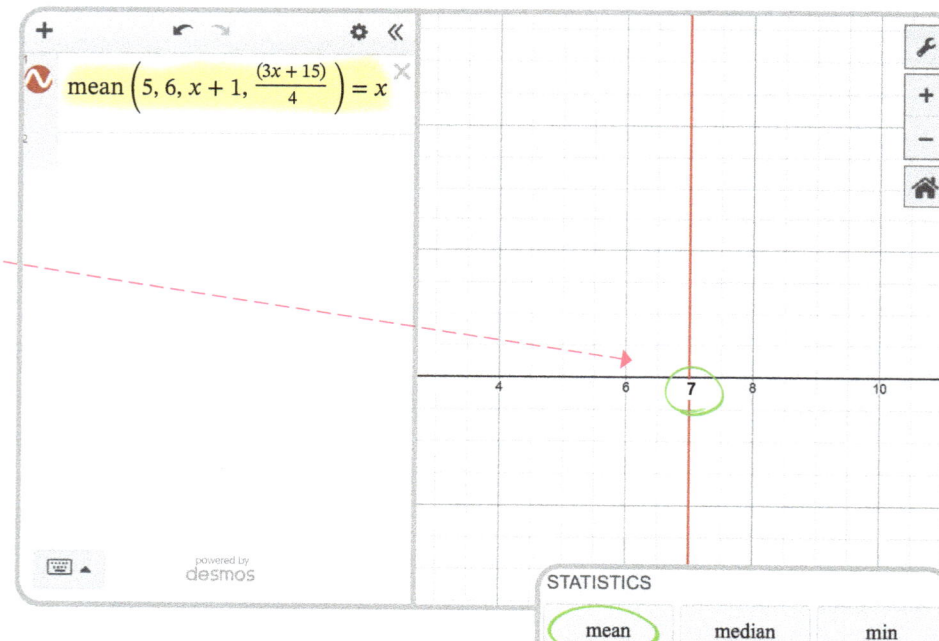

The answer is D.

Type **mean** or open functions on the Desmos Keypad and select mean.

$$4^{8d} = \sqrt[5]{4^9}$$

What is the value of d?

(A) 0.215   (B) 0.22   (C) 0.225   (D) 0.23

Note!

Replace d with x:

$$4^{8x} = \sqrt[5]{4^9}$$

How to Solve

Step 1 Enter the given equation, replacing d with x.

Step 2 Since you can't click this point, zoom in significantly to find the equation of the vertical line, $x = 0.225$.

The answer is C.

$$4^{8x} = \sqrt[5]{4^9}$$

Zoom in!

No axes? Look along the edges of the Graph Window for values.

| 0.215 | 0.22 | 0.225 | 0.23 |

NUMBER THEORY

lcm	gcd	mo
ceil	floor	rou
sign	$\sqrt[n]{\ }$	nP
nCr		

To enter $\sqrt[5]{4^9}$, either go to **functions** on the Desmos Keypad and select $\boxed{\sqrt[n]{\ }}$ or type **nthroot** and Desmos will convert it to $\boxed{\sqrt[n]{\ }}$.

Exercise #7

In triangle ABC, leg AB is 24 cm and the hypotenuse AC is 26 cm. If BC can be written as $5\sqrt{2x}$ cm, what is the value of x?

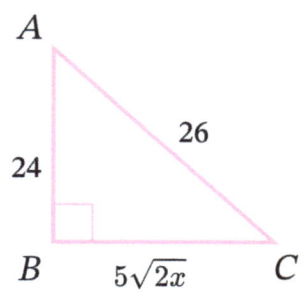

Note: Figure not drawn to scale.

(A) 2 (B) 3 (C) 4 (D) 5

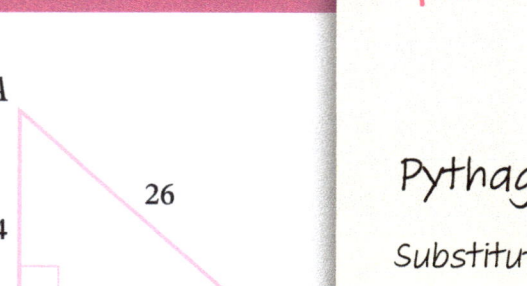

Note!

$$a^2 + b^2 = c^2$$

Pythagorean Theorem

Substitutions: $a = 24$

$b = 5\sqrt{2x}$

$c = 26$

The equation:

$$24^2 + \left(5\sqrt{2x}\right)^2 = 26^2$$

How to Solve

Step 1 Enter the equation as shown.

Step 2 Identify the x-intercept of the vertical line. The equation of the line is $x = 2$.

$$24^2 + \left(5\sqrt{2x}\right)^2 = 26^2$$

The answer is A.

Make sure the x is under the square root. Press → to exit the root before continuing.

Exercise #8

A 30−60−90 degree triangle has a perimeter of $6 + 6\sqrt{3}$ units. What is the length of the shortest side of the triangle?

(A) 2   (B) 4   (C) $2\sqrt{3}$ (D) $4\sqrt{3}$

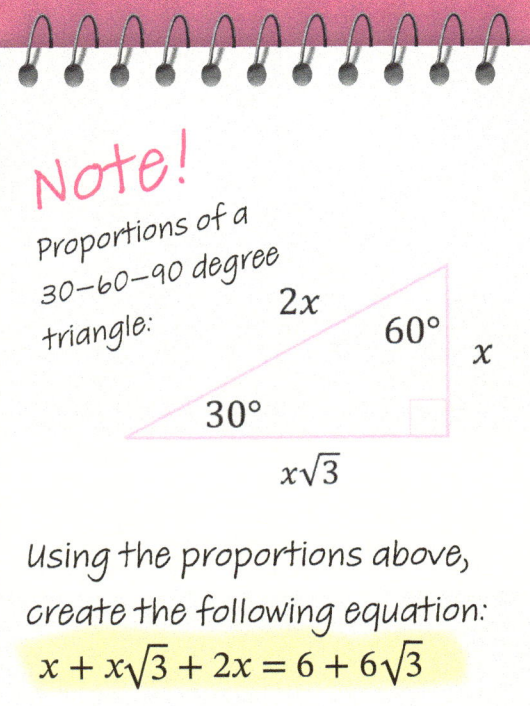

Note!

Proportions of a 30−60−90 degree triangle:

2x

60°

x

30°

$x\sqrt{3}$

Using the proportions above, create the following equation:

$$x + x\sqrt{3} + 2x = 6 + 6\sqrt{3}$$

How to Solve

Step 1 Enter the equation as shown.

Step 2 Determine the values of $2\sqrt{3}$ and $4\sqrt{3}$ by entering each expression into a row. Desmos will then display their decimal values. Find the answer choice closest to 3.464.

The answer is C.

$x + x\sqrt{3} + 2x = 6 + 6\sqrt{3}$

$2\sqrt{3}$

= 3.46410161514

$4\sqrt{3}$

= 6.92820323028

(3.464,0)

$2\sqrt{3} \sim 3.464$

powered by desmos

On the SAT exam, the proportions of special triangles are provided in the reference section.

Technique #1
PRACTICE PROBLEMS

Problem 1

What is one possible solution to the equation $(x + 1)^2 - 5(x + 1) + 6 = 0$?

Problem 2

What percent of 300 is 750?

A) 250%
B) 275%
C) 300%
D) 325%

Hint: Write the equation like this:
$x\% \text{ of } 300 = 750$

Problem 3

For what value of x does
$x^2 - 6x + 9 = (x + 3)^2$?

Problem 4

$x + 3 = \sqrt{x + 5}$

What is the solution set of the equation above?

A) $\{-4\}$
B) $\{-1, -3\}$
C) $\{-1\}$
D) $\{-1, -4\}$

Technique #1
PRACTICE PROBLEMS

Problem 5

In a right triangle, one leg is 3 units and another leg is $6\sqrt{x}$ units. The hypotenuse is $3\sqrt{29}$ units. What is x?

A) 7
B) 49
C) 78
D) $\sqrt{78}$

Problem 6

If $|x + 5| = 6$ and x is a negative number, what is the value of x?

Problem 7

$$\frac{x - 100}{2}, \frac{x}{5}, \frac{x + 50}{3}$$

The average (arithmetic mean) of the data above is 75. What is the value of x?

A) 150
B) 200
C) 250
D) 300

Problem 8

$$2^{2x} + 3^{4x} = 100$$

What is the value of x, rounded to the nearest hundredth?

A) 1.01
B) 1.02
C) 1.03
D) 1.04

Hint: Zoom in!

Problem 9

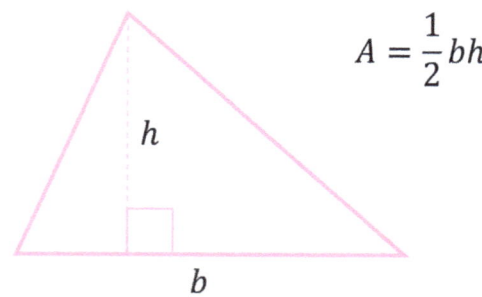

$$A = \frac{1}{2} bh$$

Note: Figure not drawn to scale.

A triangle has a base of $2\sqrt{3x}$ units and a height of $3\sqrt{3x}$ units. If the area is 99 square units, what is the value of x?

_____

Problem 10

A 45−45−90 degree triangle has a perimeter of $24 + 24\sqrt{2}$ units. What is the length of the shortest side of the triangle?

A) 7
B) $12\sqrt{2}$
C) 78
D) $24\sqrt{2}$

Proportions of a 45−45−90 degree triangle:

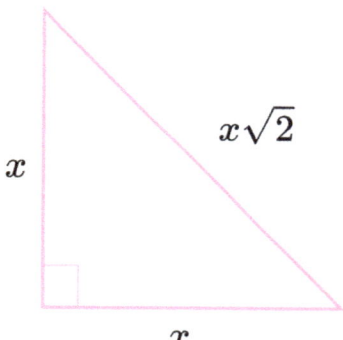

Note: Later geometry problems won't include these proportions, so refer to page 17 whenever you need the full list.

Technique #1
PRACTICE PROBLEMS

Answers		References
1	1 or 2	*Exercise #1*
2	A	*Exercise #4*
3	0	*Exercise #1*
4	C	*Exercise #2*
5	A	*Exercise #7*
6	−11	*Exercise #3*
7	C	*Exercise #5*
8	D	*Exercise #6*
9	11	*Exercise #7*
10	B	*Exercise #8*

Notes

Technique

2

Coordinate Points:
Tiny but Mighty

Coordinate Points: Tiny but Mighty

Technique #2 focuses on coordinate points, which are the building blocks of equations and graphs. These points may seem small, but they often carry important information in a problem.

In this technique, you'll work with key points such as intercepts, vertices, and maximum or minimum values. You'll also practice using tables, sliders, and different graph types.

This technique covers a wide range of problem types, helping you recognize how coordinate points play different roles across problems.

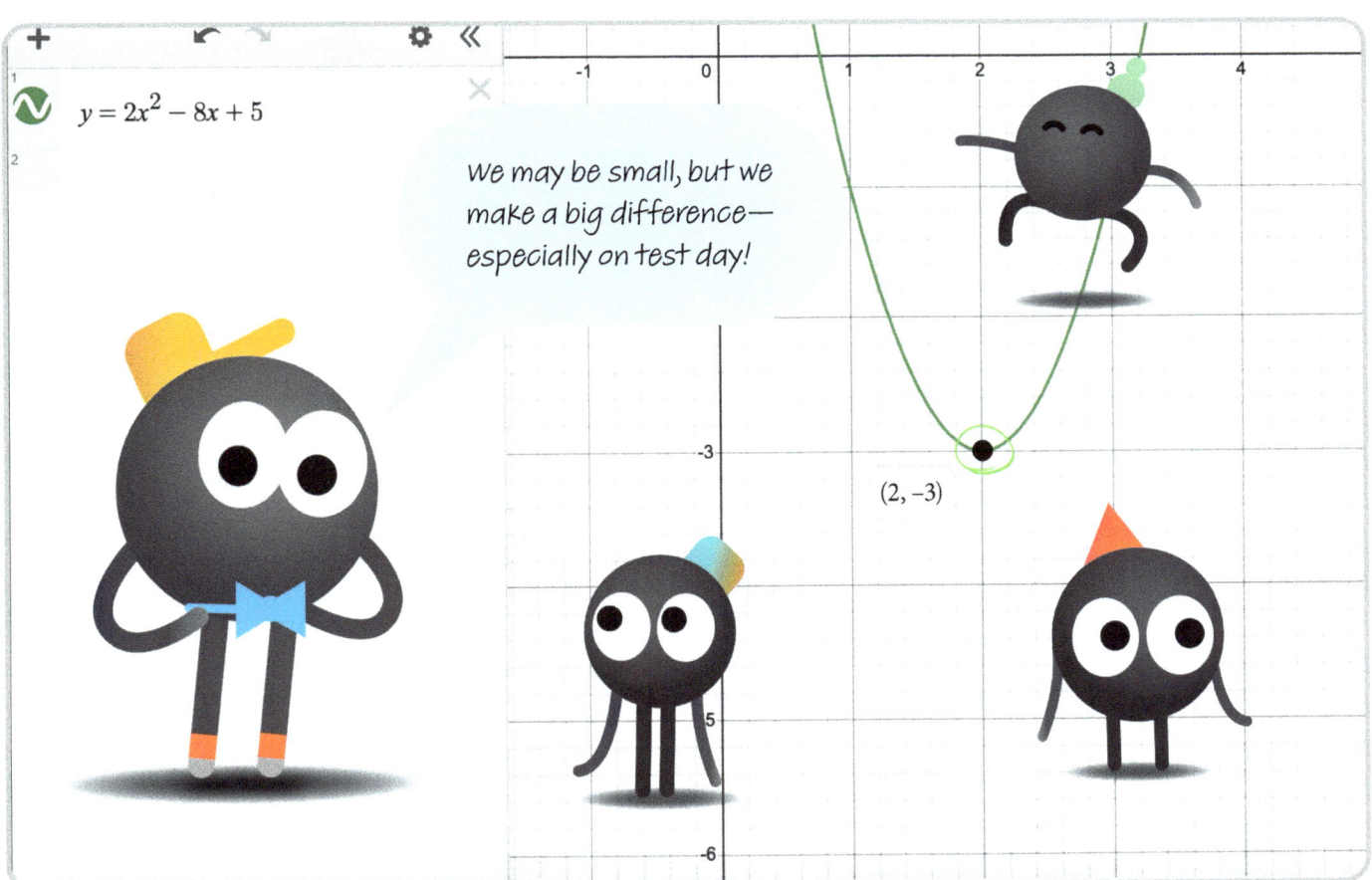

Exercise #9

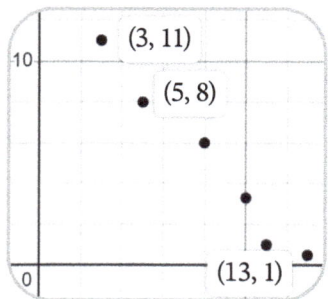

Which equation best represents a linear model for the data on the scatter plot above?

A) $x + 2y = 26$ B) $x + y = 13$
C) $2x + y = 13$ D) $3x + y = -13$

Note!

Choose points that follow the overall trend of the scatter plot. Here, all the points fit the trend, so suitable points include (3, 11), (5, 8), and (13, 1).

How to Solve

Step 1 Enter the points, separating each point with a comma.

Step 2 Enter each answer choice in a separate row.

Step 3 Identify the line that closely follows the coordinate points from Step 1.

The answer is B.

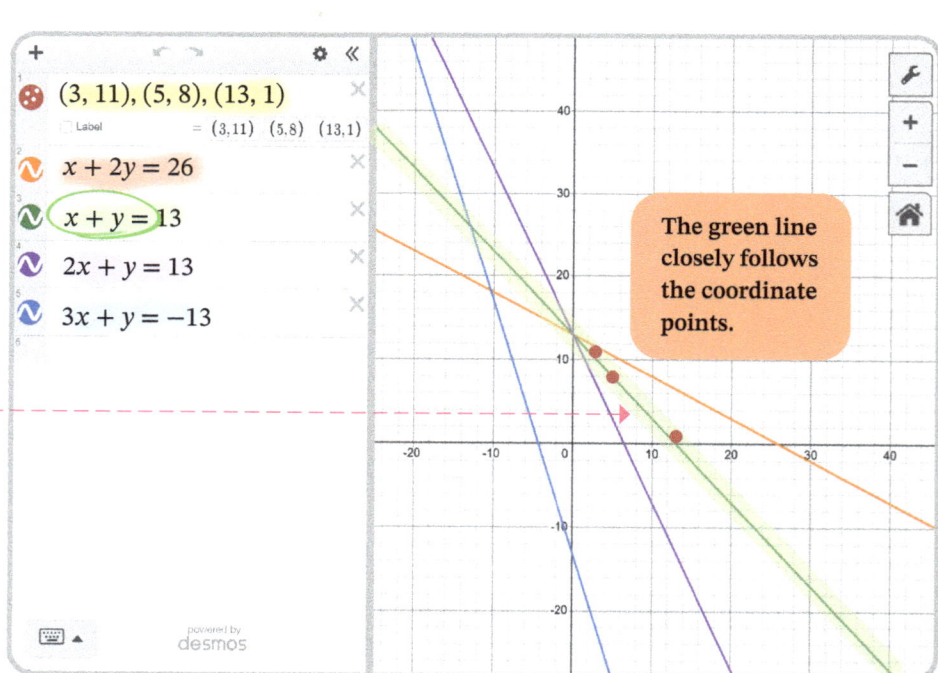

The green line closely follows the coordinate points.

Exercise #10

$h(x) = 2x^4 + 3x^3 - 7x^2 + 20$

Which table shows correct values for $h(x)$?

Note!

Desmos can generate a Function Table for equations that have y (or the function) isolated.

A)

x	1	2	3
$h(x)$	18	20	200

B)

x	1	2	3
$h(x)$	0	200	612

C)

x	1	2	3
$h(x)$	20	48	200

D)

x	1	2	3
$h(x)$	18	48	200

How to Solve

Edit List

Step 1 Enter the given equation and select **Edit List** ⚙.

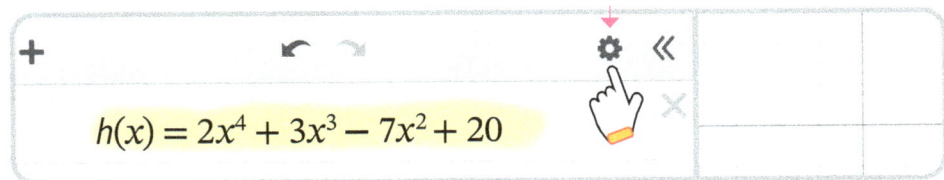

Step 2 Click **Create Table** ▦. A Function Table will automatically appear, showing the values that satisfy the given equation.

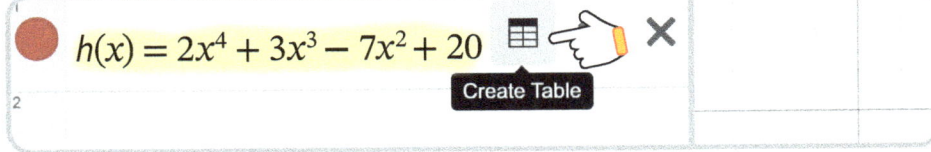

Create Table

Step 3 Add an x value of 3 to the table, and the corresponding $h(x)$ value will appear. Find the answer choice that matches the coordinate points in the table.

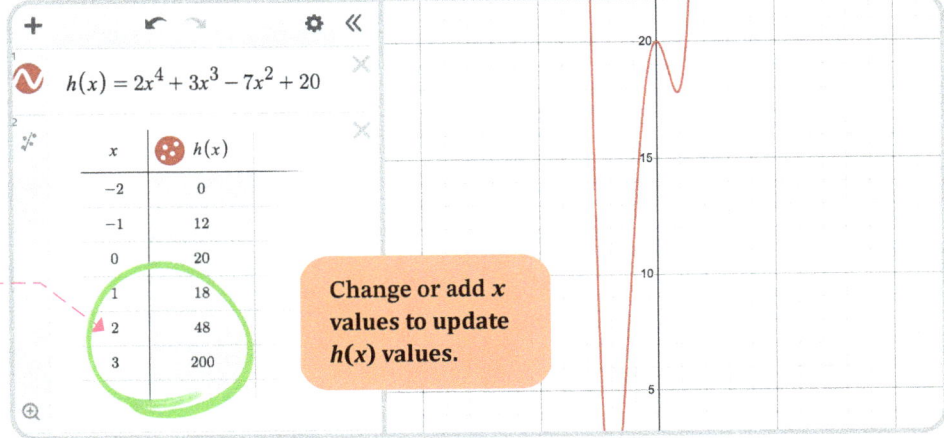

Change or add x values to update $h(x)$ values.

The answer is D.

Desmos offers a slider when you start writing a function, but don't accept it.

Note!

Data points in the Function Table help tell the story of the equation.

The equation $p(x) = 800(0.92)^x$ gives the estimated population of a bird species in a wildlife area, where x represents the number of years since tracking began. What does the number 800 represent?

(A) The percent increase in population each year
(B) The bird species' population when tracking began
(C) The percent decrease in population each year
(D) The number of years since tracking began

How to Solve

Edit List

Step 1 Enter the given equation and select **Edit List** ⚙.

$p(x) = 800(0.92)^x$

x	$p(x)$
-2	945.17958
-1	869.56522
0	800
1	736
2	677.12

powered by
desmos

Step 2 Click **Create Table** 🄳. The value 800 appears when $x = 0$, which means there were 800 birds at the start of the tracking period.

Use the table to interpret the value of 800.

The answer is B.

Tried counting the birds myself. They did NOT appreciate it.

Exercise #12

$$x^2 - 6x + y^2 - 8y = 11$$

What is the diameter of a circle represented by the equation above?

(A) 6   (B) 18   (C) 12   (D) 7

Note!

If the circle appears distorted, click **Default View** 🏠 to reset the axes.

How to Solve

Step 1 Enter the equation.

Step 2 Click the minimum and maximum points to reveal their coordinates.

Step 3 To find the diameter, subtract the minimum y value from the maximum y value.

The answer is C.

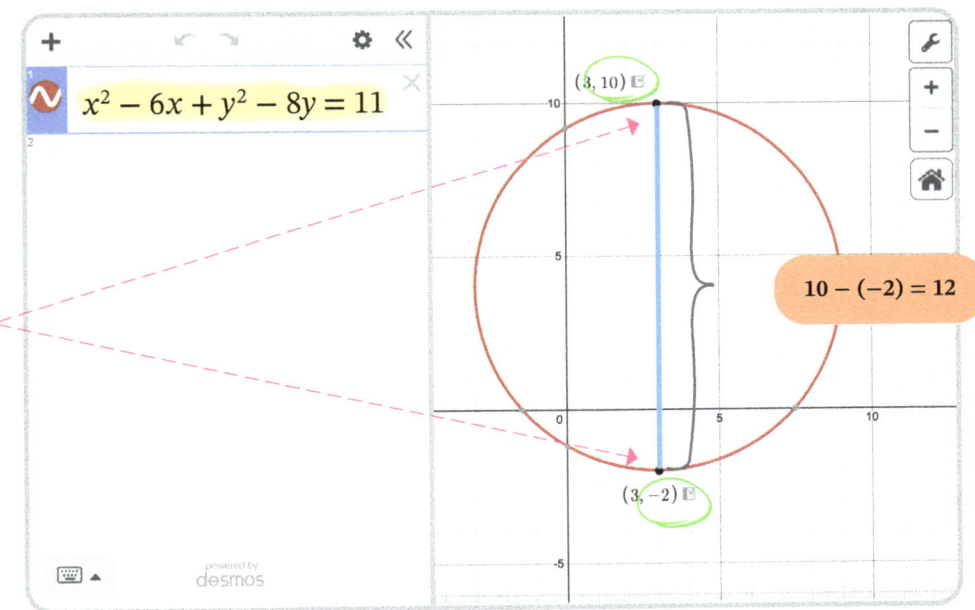

To find the diameter, you can also type the following in the Expression List:

distance $((3, 10),(3, -2))$

$= 12$

Exercise #13

A circle has the equation
$x^2 - ax + y^2 - 16y = -16$, where a is a constant.
The diameter of the circle has endpoints $(4, 0)$
and $(4, 16)$. What is the value of a?

(A) 8　(B) −8　(C) 6　(D) −6

Note!

Desmos creates a slider for an unknown constant because it needs a value for it to graph.

How to Solve

Step 1 Enter the points.

Step 2 Enter the given equation and add a slider for a.

Step 3 Adjust the slider until the circle passes through the given endpoints.

> The answer is A.

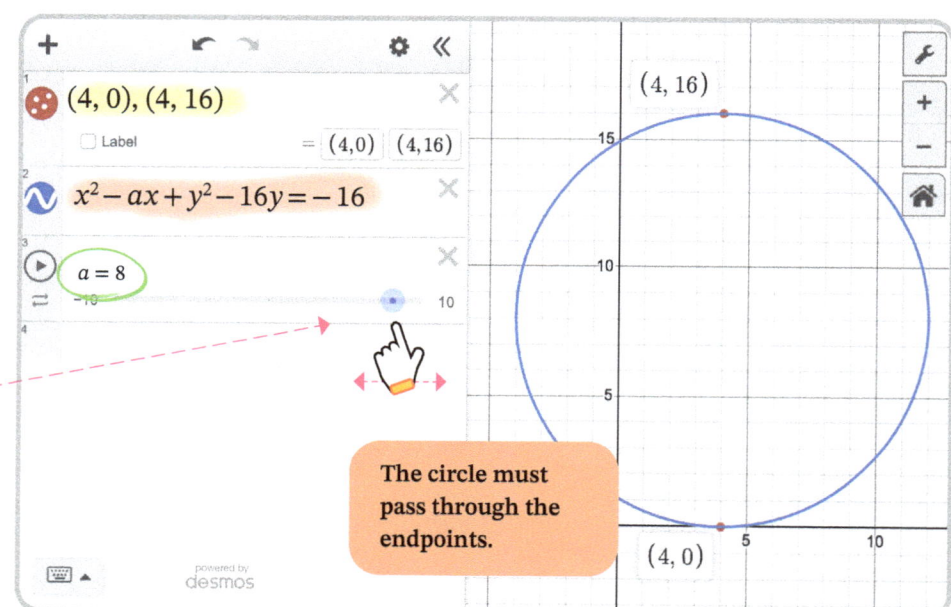

The circle must pass through the endpoints.

Sometimes the slider option disappears, so it's easiest to accept the slider as soon as the prompt appears.

Exercise #14

The graph of a line passes through $(-4, k - 4)$, $(8, k + 8)$, and $(0, 2)$, where k is a constant. What is one possible value of k?

(A) 8   (B) 6   (C) 4   (D) 2

How to Solve

Step 1 Enter the points, separated by commas. Add a slider for k.

Step 2 Turn **Lines** on. (See note below.)

Step 3 Adjust the slider to find the value that creates a line between all three points.

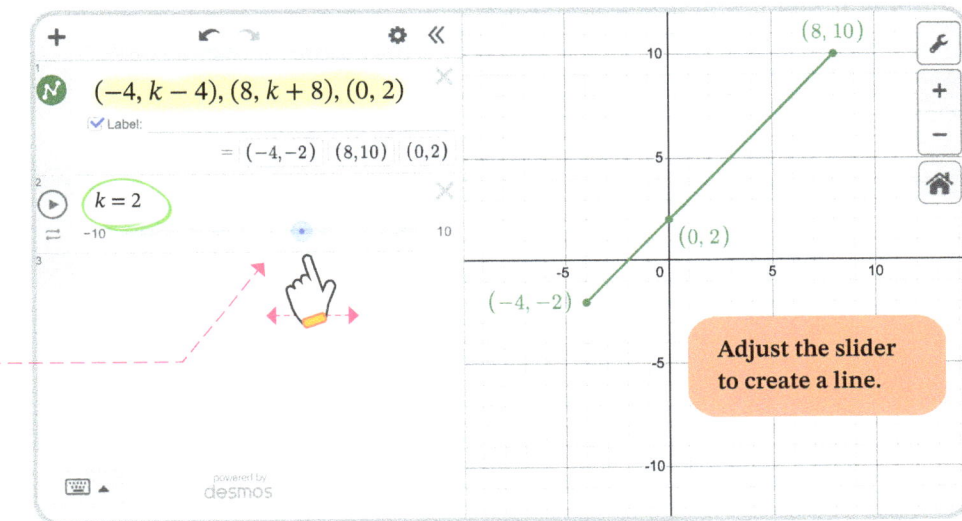

Adjust the slider to create a line.

The answer is D.

Connecting points

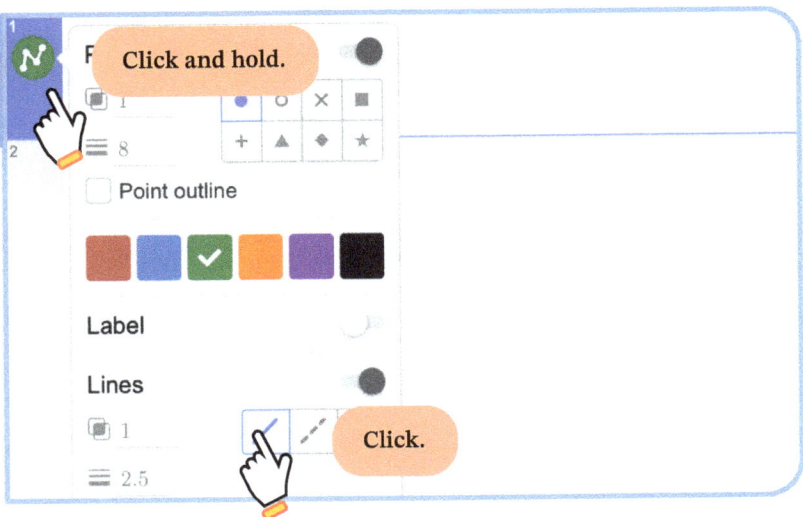

Click and hold.

Point outline

Label

Lines

Click.

To connect points, enter all coordinate points in the same row, separated by commas. Then click and hold the **Color Icon**  to open the Color Panel and turn on **Lines**.

A quadratic function has a vertex at $(-9, 75)$ and passes through $(-4, 50)$. The equation of the function is $y = -x^2 + bx - 6$. What is the value of b?

_____

Note!

Since this equation includes an unknown constant, a slider is needed for b.

How to Solve

Step 1 Enter and label the given points.

Step 2 Enter the given equation and add a slider for b.

Step 3 Adjust the slider and its range until the parabola passes through $(-14, 50)$ and has a vertex at $(-9, 75)$.

The answer is -18.

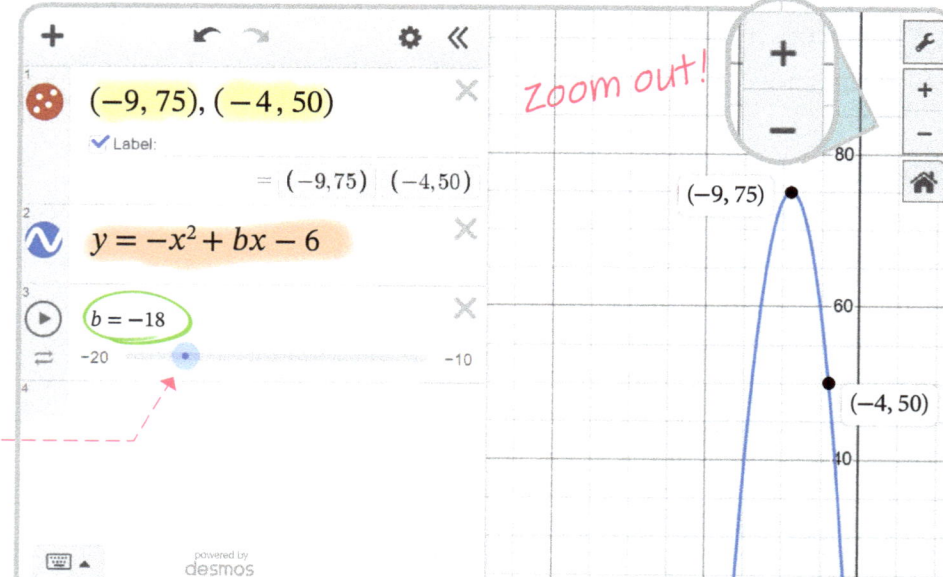

To change the slider's range, click in the slider's row and enter new values:

$-20 \leq b \leq -10$

Then click outside the row.

Exercise #16

Note!

When graphing functions, use the variable x. Replace n with x:

$$f(x) = 2x^2 + 12x - 224$$

$f(n) = 2n^2 + 12n - 224$

The given equation defines the function f. For what value of n does $f(n)$ reach its minimum?

_____
_____

How to Solve

Step 1 Enter the given equation as shown.

Step 2 Click the vertex to find the x value.

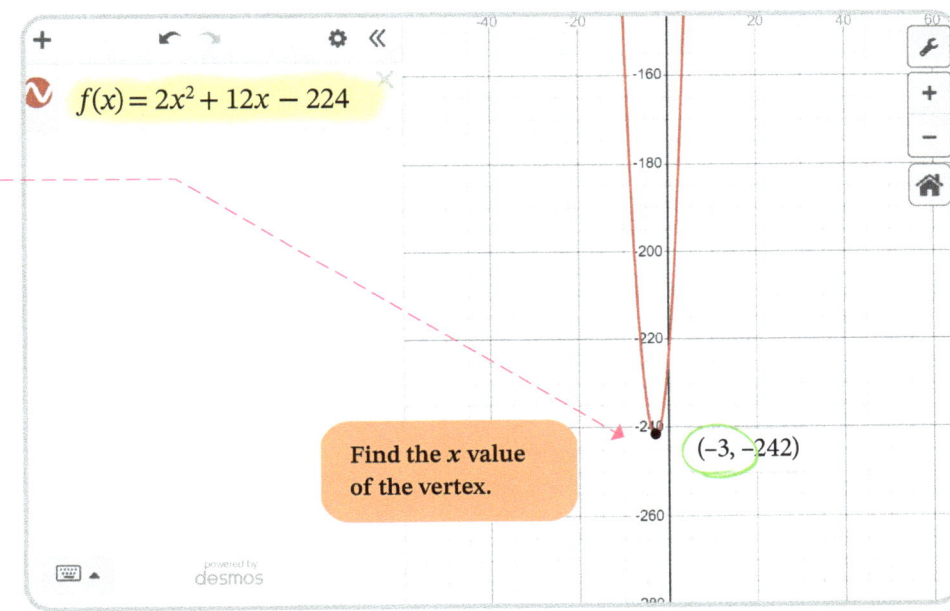

$f(x) = 2x^2 + 12x - 224$

Find the x value of the vertex.

$(-3, -242)$

The answer is −3.

powered by desmos

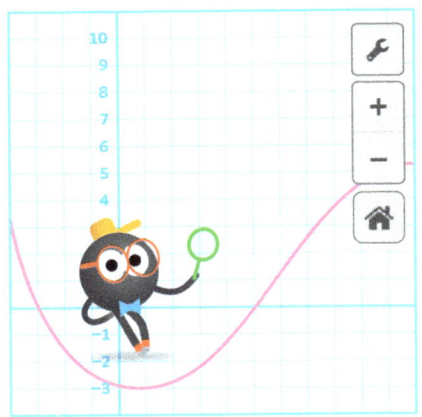

Navigating graphs is easier when you become friends with the zoom tools.

Exercise #17

$f(x) = -2x^2 + 2x + 8$

The equation above defines the function f. If $g(m) = f(m - 2)$, what is $g(m)$ at its maximum point?

Note!

Rewrite the second function to replace m with x:

$g(x) = f(x - 2)$

(A) −4   (B) 0.5   (C) 2   (D) 8.5

How to Solve

Step 1 Enter the given equations.

Step 2 Click the vertex of $g(x)$ to find the y value.

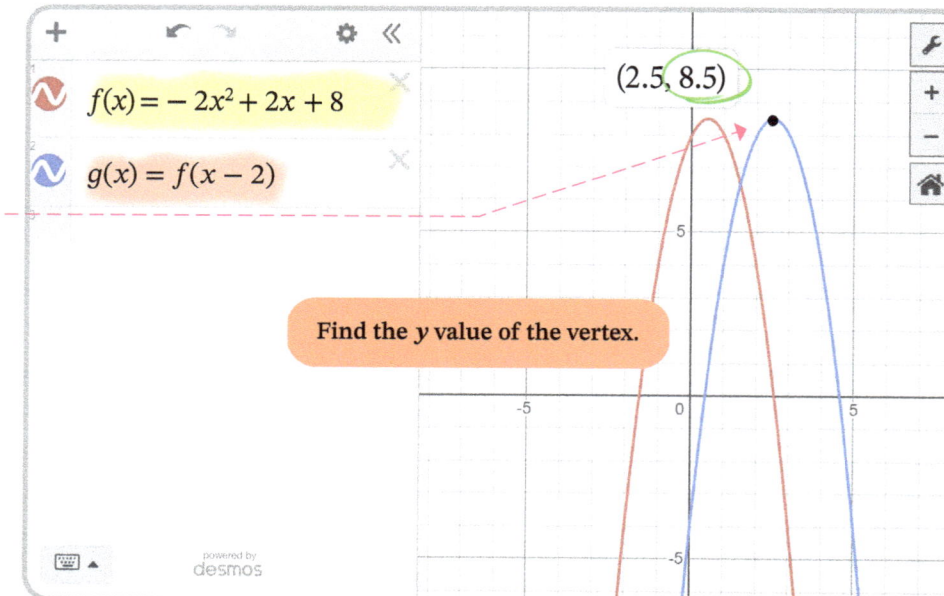

$f(x) = -2x^2 + 2x + 8$

$g(x) = f(x - 2)$

Find the y value of the vertex.

$(2.5, 8.5)$

The answer is D.

powered by desmos

The answer to $g(m)$ is a y value. If the question had asked for the value of m, the answer would have been an x value.

Exercise #18

A paper rocket is launched from 12 meters above the ground. Its height (in meters) is given by the equation $y = n - (3x - m)^2$, where x represents the time (in seconds) after launch and m and n are constants. The rocket reaches its maximum height of 93 meters when $x = 3$. What is its height 2 seconds after launch?

Note!

Use these points.

Initial height: $(0, 12)$

Maximum height: $(3, 93)$

_____

How to Solve

Step 1 Enter and label the points. Then enter the given equation.

Step 2 Accept the sliders for m and n and adjust them until the graph passes through $(0, 12)$ and reaches its maximum height at $(3, 93)$.

Step 3 Click and drag along the parabola to reach $(2, 84)$.

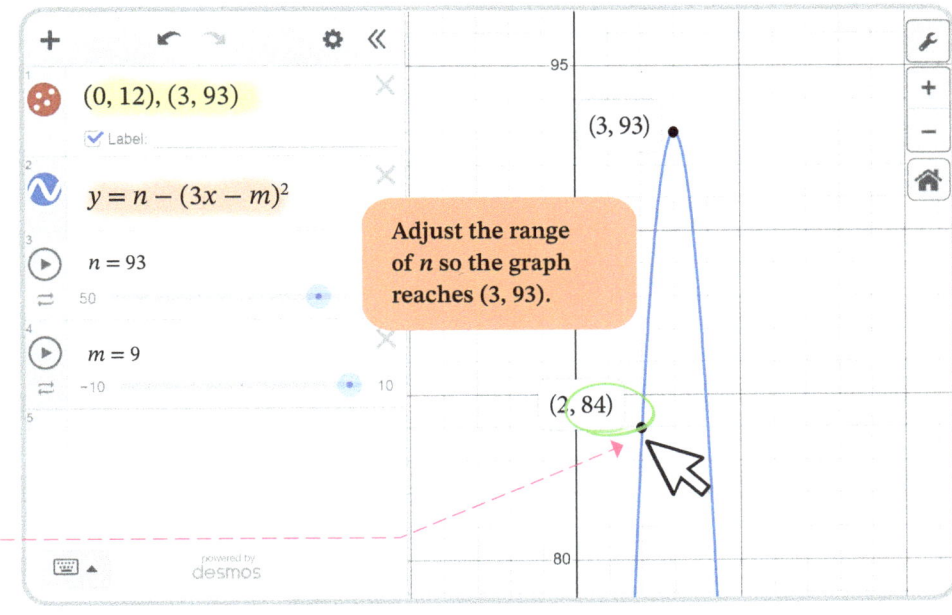

$(0, 12), (3, 93)$
☑ Label:

$y = n - (3x - m)^2$

$n = 93$
50

$m = 9$
−10   10

Adjust the range of n so the graph reaches $(3, 93)$.

$(3, 93)$

$(2, 84)$

95

80

powered by desmos

The answer is 84.

$(2, 84)$

For Step 3, you can also type $x = 2$ in a new row and find the intersection of the line and parabola.

Technique #2
PRACTICE PROBLEMS

Problem 1

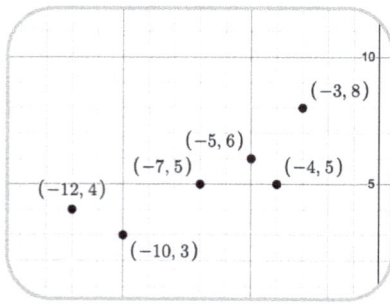

Which of the following equations best fits the data shown in the scatter plot above?

A) $3x + 7y = 56$
B) $6x - 14y = 112$
C) $-3x = 144 - 7y$
D) $-7x + 10y = 100$

Problem 2

A line passes through $(k, 4)$, $(9, k)$, and $(0, 0)$. Which of the following could be the value of k?

A) -2
B) -4
C) -6
D) -8

Problem 3

$$f(x) = 2(x - 1)^2 + 3$$

The given equation defines the function f. If $g(x) = f(x + 4) + 5$, what are the coordinates of the vertex of $g(x)$?

A) $(-4, 3)$
B) $(-3, 8)$
C) $(-2, 11)$
D) $(-1, 3)$

Problem 4

A quadratic function passes through $(-2, 62)$ and $(5, 20)$. If its equation is $y = 4x^2 - bx + 10$, what is the value of b?

Hint: Expand the range for b.

Technique #2
PRACTICE PROBLEMS

Problem 5

The graph of an exponential equation passes through the point $(1, m)$, where m is a constant. Which of the following equivalent equations shows the value of m in the equation itself?

A) $y = 45(1.3)^{x+1}$
B) $y = 58.5(1.3)^{x}$
C) $y = 76.05(1.3)^{x-1}$
D) $y = 98.97(1.3)^{x-2}$

Hint: Create a Function Table for each equation and look for the y value when $x = 1$. Identify which equation includes that y value (the value of m) in the equation itself. This is a challenging problem, but it's realistic—it's based on a question from an official SAT practice test.

Problem 6

A water balloon is tossed upward from a balcony. Its height in feet, y, can be determined by the equation $y = -16x^2 + 24x + 40$, where x represents the time in seconds after the balloon is thrown. Which of the following statements is true about the water balloon?

A) The water balloon was thrown from a height of 49 feet.

B) The water balloon reached a maximum height of 49 feet.

C) It took 49 seconds for the water balloon to reach its maximum height.

D) It took 49 seconds for the water balloon to hit the ground.

Technique #2
PRACTICE PROBLEMS

Problem 7

A toy car was launched from a height of 80 meters. Its height at time t is given by the function $h(t) = a - (4t - b)^2$.

If the toy car's maximum height of 84 meters occurred 0.5 seconds after it was launched, what was the height of the toy car 2 seconds after launch?

A) 10 meters
B) 24 meters
C) 48 meters
D) 80 meters

Problem 8

A circle's diameter has endpoints $(-2, -11)$ and $(8, -11)$. The equation of the circle is $(x - 3)^2 + (y + 11)^2 = c$. What is the value of c?

_____

Problem 9

Given the function $f(x) = x^2 + 6x + 7$, which table shows three x values and their corresponding $f(x)$ values?

A)

x	$f(x)$
−2	−1
0	7
2	23

B)

x	$f(x)$
−3	10
0	7
3	22

C)

x	$f(x)$
−1	12
1	14
3	22

D)

x	$f(x)$
−4	7
1	14
3	22

Problem 10

The equation of a circle is given as $x^2 + 6x + y^2 - 14y = 42$. What is the radius of the circle?

A) 10
B) 20
C) 5
D) $\sqrt{10}$

Answers

1 D

2 C

3 B

4 18

5 C

6 B

7 C

8 25

9 A

10 A

References

Exercise #9

Exercise #14

Exercise #17

Exercise #15

Exercise #11

Exercise #16

Exercise #18

Exercise #13

Exercise #10

Exercise #12

Notes

Technique

#

The Power of
Agreement

The Power of Agreement

Technique #3 focuses on systems, which are sets of two or more equations or inequalities that share variables.

In a system, a solution occurs where graphs agree—where they intersect. Instead of solving equations one variable at a time, you can often spot solutions visually by identifying points where the graphs meet.

In this technique, you'll use intersections to solve a wide range of SAT questions, including those involving compound inequalities, absolute values, and functions.

Want the solution? Follow the graphs to where they agree. We'll meet you there.

$y = \frac{1}{2}x + 1$

$x + 2y = 10$

The solution to the given system of equations is (x, y). What is the value of x?

(A) 4   (B) 0.5   (C) 1   (D) 3

Note!

The solution to a system of equations is the point where the graphs intersect.

How to Solve

Step 1 Enter the two equations.

Step 2 Click the intersection to find the x value.

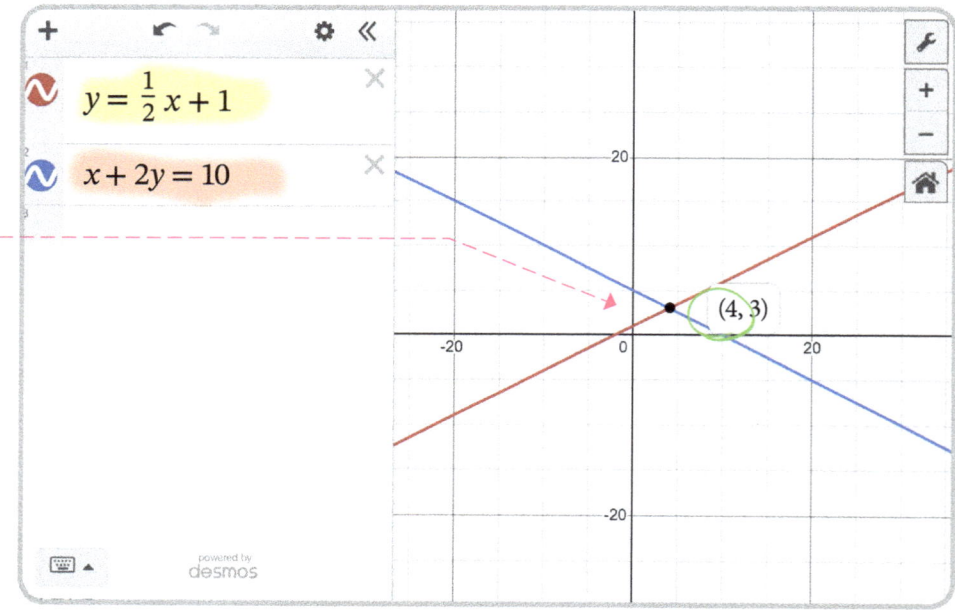

$y = \frac{1}{2}x + 1$

$x + 2y = 10$

$(4, 3)$

powered by desmos

The answer is A.

Some points are like diplomats—helping graphs connect and find common ground.

You're both lines. That's a promising start.

Exercise #20

The function g is defined by $g(x) = -9x^2 + 6x + 25$. If $g(x) = 26$, what is the value of $3x$?

Note!

Because Desmos shortens repeating decimals, use the fraction form or round your answer to the fourth decimal place.

How to Solve

Step 1 Enter the given equations.

Step 2 Click the intersection to find the x value.

Step 3 Multiply 0.33333 by 3 to get 0.99999, then round to 1. Alternatively, use $\frac{1}{3}$ instead of 0.33333, and multiply by 3 to get 1.

The answer is 1.

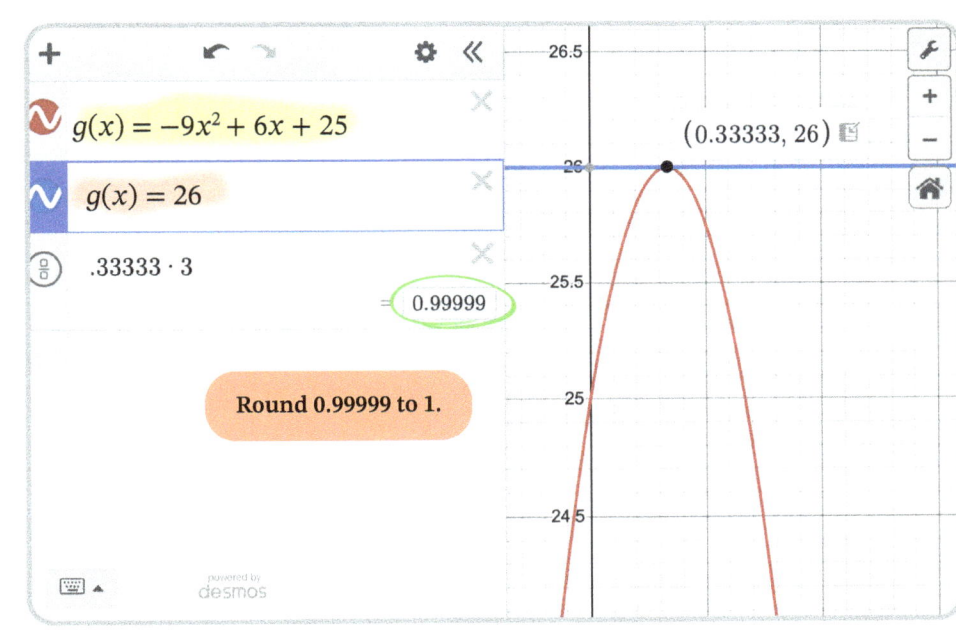

Hmmm... 0.33333? That's just $\frac{1}{3}$ in disguise! For common equivalencies, see **Truncated Decimals** in the glossary.

Note!

when graphing a function, use *x* as the variable. Replace *d* with *x* to create the following:

$$3f\left(\frac{x}{2}\right) = 36$$

The function f is defined by $f(x) = x + 4$. If $3f\left(\frac{d}{2}\right) = 36$, what is the value of d?

(A) 4   B) 6   (C) 16   (D) 20

How to Solve

Step 1 Enter the equations as shown.

Step 2 Click the intersection to find the *x* value.

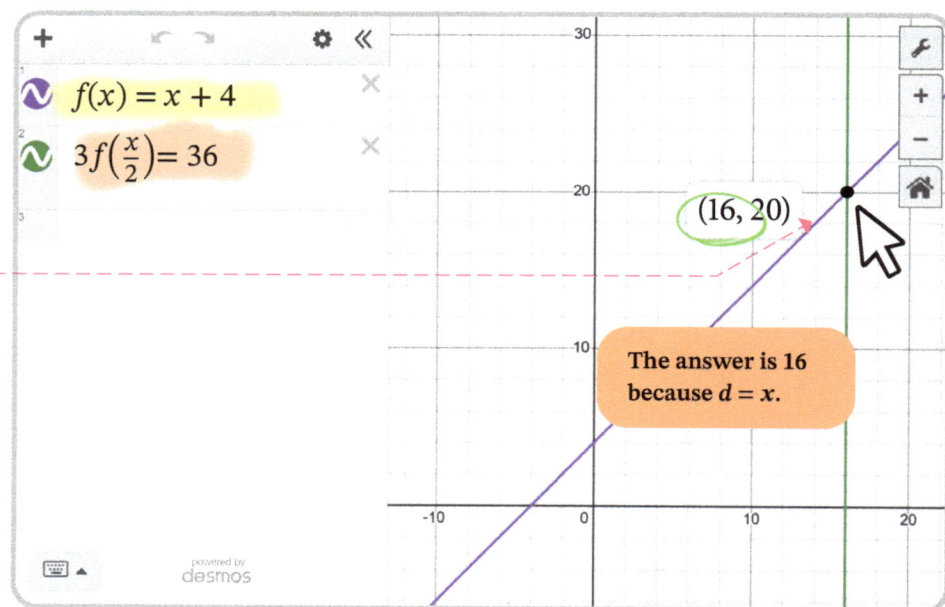

$f(x) = x + 4$

$3f\left(\frac{x}{2}\right) = 36$

(16, 20)

The answer is 16 because $d = x$.

powered by desmos

The answer is C.

To hit the target, be like a function—focused, steady, and predictable.

Exercise #22

Cafe Barone can seat up to 72 people. The cafe has 26 tables—some tables can seat up to 2 guests, while others can seat up to 4 guests. How many tables can seat 4 people?

_____

Note!

of 2-seater tables: x
# of 4-seater tables: y

Equations
Total tables: $x + y = 26$
Maximum people seated:
$2x + 4y = 72$

How to Solve

Step 1 Enter the equations as shown.

Step 2 Click the intersection to find the y value.

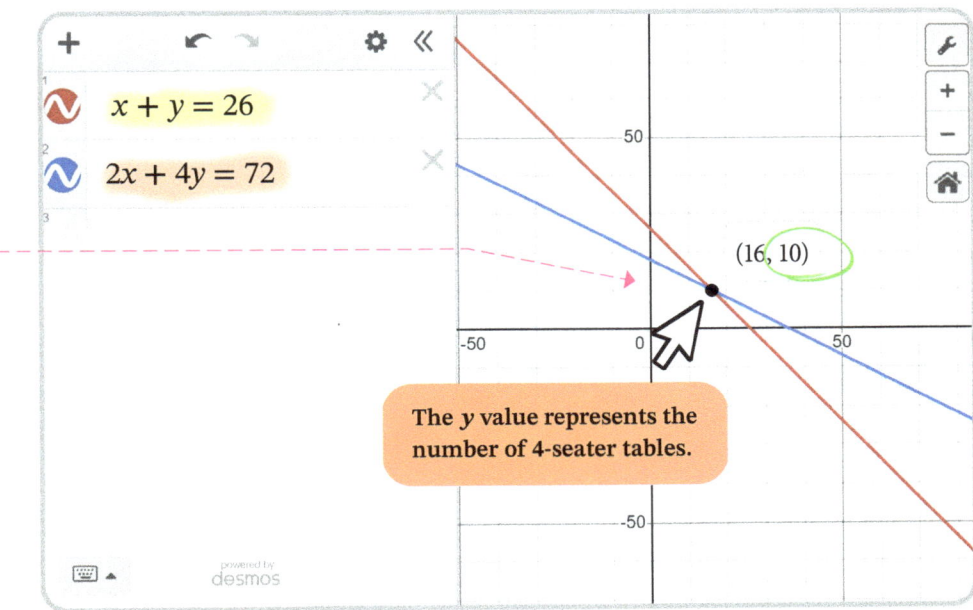

$x + y = 26$

$2x + 4y = 72$

$(16, 10)$

The y value represents the number of 4-seater tables.

The answer is 10.

Once you create the equations, let Desmos find the solution.

Desmos doesn't offer a slider for a because it treats it like x. To fix this, type $a = 5$ (or any number) in a new row, then click outside the row to activate the slider.

$y = -x^2 + 24x - 150$

$y = a$

In the given system of equations, a is a constant. What value of a results in exactly one solution for the system of equations?

_____

How to Solve

Step 1 Enter the given equations. Since Desmos does not offer a slider for a, type $a = 5$ in the next row to prompt a slider.

Step 2 Adjust the slider until the parabola intersects the line at exactly one point.

The answer is −6.

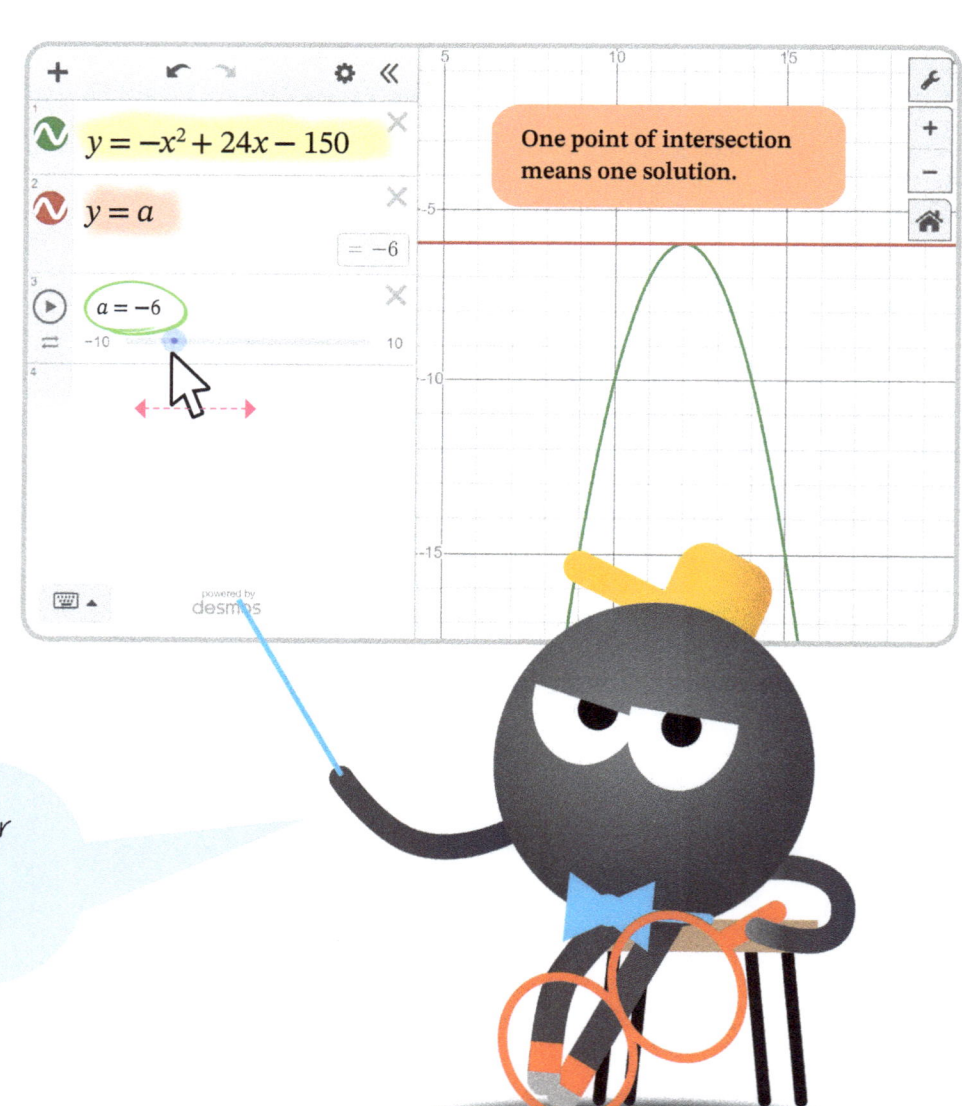

One point of intersection means one solution.

If you don't add a slider for a, Desmos will incorrectly graph $y = a$ as $y = x$.

Exercise #24

$y > 9$

$3x + y \leq 1$

Which of the following points is a solution to the system of inequalities?

(A) $(-20, 40)$ (B) $(-6, 40)$

(C) $(8, 40)$ (D) $(26, 40)$

Note!

Solutions to inequalities are found in the double-shaded region.

How to Solve

Step 1 Enter the given inequalities.

Step 2 Enter each coordinate point from the answer choices and select **Label**.

Step 3 Identify the point that lies within the double-shaded region.

The answer is A.

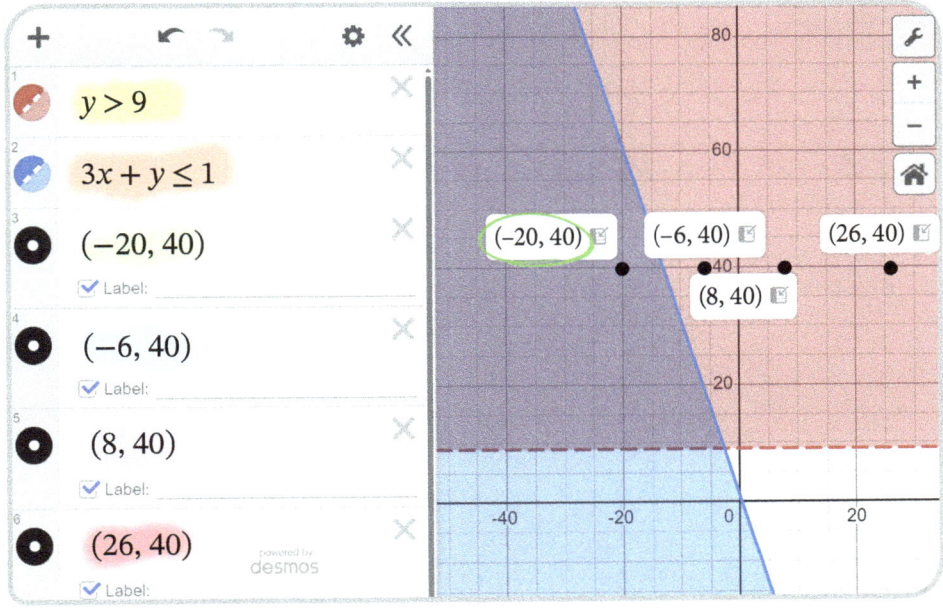

Double shading? Must be important.

Exercise #25

For some positive integer x, $4 < |2x - 5| < 6$. What is the value of x?

_____

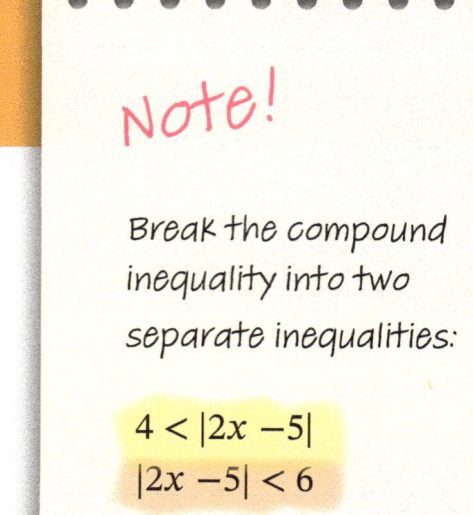

Note!

Break the compound inequality into two separate inequalities:

$4 < |2x - 5|$
$|2x - 5| < 6$

How to Solve

Step 1 Write the inequalities as shown.

Step 2 Find the positive integer value of x in the double-shaded region (purple).

$4 < |2x - 5|$

$|2x - 5| < 6$

5 is the only positive integer in the double-shaded region.

The answer is 5.

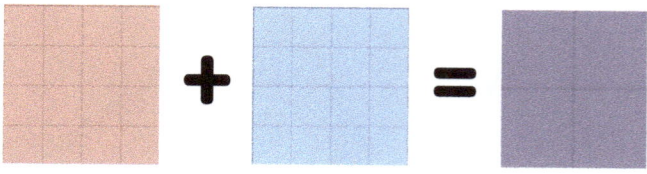

The color of the double-shaded region is a blend of the colors used for the individually shaded areas.

Exercise #26

$50x^2 + 35x = y + 100$

$49x^2 = 300 + y + 24x$

The solution to the given system of equations is (x, y). Which of the following is closest to one possible value of x?

(A) -3.52 (B) -3.61

(C) -3.63 (D) -3.67

warning!!

Technique #3 looks like it should work here, but the intersection is hard to spot. Using vertical lines instead (Technique #1) gives you a clearer answer.

How to Solve

Step 1 Isolate y in both equations:
$y = 50x^2 + 35x - 100$
$y = 49x^2 - 24x - 300$

That's a long equation! Adjust the length of the Expression List to fit it.

Step 2 Set the right sides of the equations equal to each other, since both represent y:
$50x^2 + 35x - 100 = 49x^2 - 24x - 300$

Step 3 Enter the combined equation.

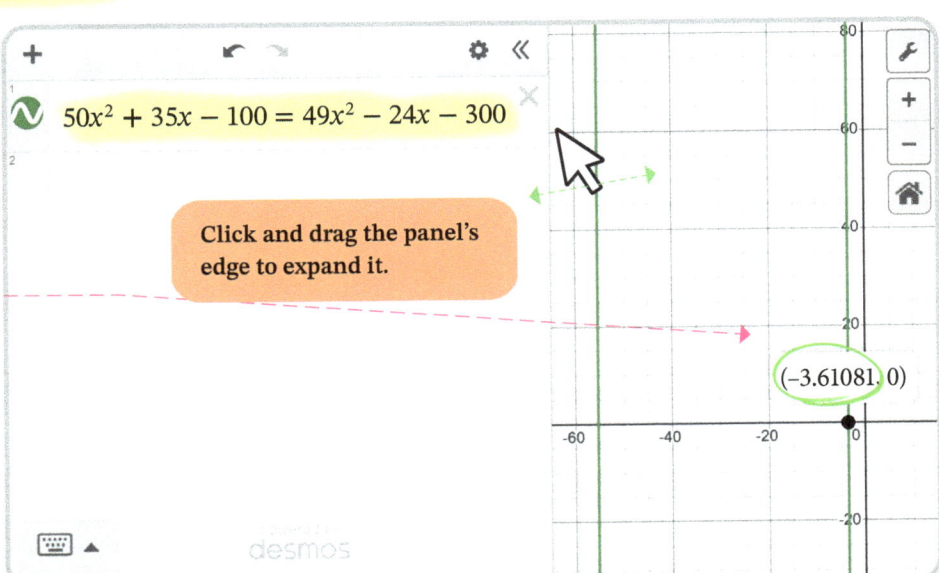

$50x^2 + 35x - 100 = 49x^2 - 24x - 300$

Click and drag the panel's edge to expand it.

$(-3.61081, 0)$

Step 4 Identify the answer choice closest to the x value of one of the vertical lines.

The answer is B.

desmos

Technique #3
PRACTICE PROBLEMS

Problem 1

$x + 10y = 11$

$2x - y = 1$

The solution to the following system of equations is (a, b). What is the value of b?

A) 1

B) 2

C) 3

D) 4

Problem 2

Let the function h be defined by $h(x) = 3x - 3$. If $\frac{h(k)}{3} = 36$, what is the value of k?

A) 32

B) 34

C) 36

D) 37

Problem 3

At Segoshi's Japanese restaurant, each sushi roll contains either 6 or 8 pieces. If Ella ordered 9 sushi rolls containing a total of 62 pieces, how many 8-piece rolls did she order?

A) 30

B) 32

C) 4

D) 5

Problem 4

$y = 3$

$y = x^2 + 10x + n$

In the given system of equations, n is a positive constant. If the system has exactly one solution, what is the value of n?

_____

Hint: Desmos gives you a slider for $y = 3$, but don't adjust it. Instead, adjust the slider for n.

Technique #3
PRACTICE PROBLEMS

Problem 5

For some negative integer x, $8 < |2x - 3| < 10$. What is the value of x?

_____

Problem 6

$600x^2 - 159 = 2y$
$y = 200x^2$

The solution to the given system of equations is (x, y). What is the closest value of x if $x > 0$?

A) 0.875
B) 0.879
C) 0.892
D) 0.902

Hint: Combine the two equations to get $600x^2 - 159 = 2(200x^2)$.

Problem 7

$y = d$
$y = x^2 - 8x - 4$

In the system above, d is a constant and an integer. If the system has no solutions, what is the greatest possible value of d?

_____

Problem 8

$g(x) = 6x^2 + 7x - 10$

Function g is shown above. If $g(x) = 0$, what is the positive value of $6x$?

_____

Hint: Change 0.83333 to $\frac{5}{6}$ before multiplying by 6. Alternatively, round the answer.

Problem 9

$y > 5$

$2x + y < 3$

Which of the following points is a solution to the system of inequalities above?

A) $(0, 6)$

B) $(-1, 6)$

C) $(4, 5)$

D) $(-4, 7)$

Problem 10

$h(x) = 2x^2 - 8x - 20$

The function h is defined above. If $h(2x) = 100$, what is one possible value of x?

A) 35

B) 25

C) 15

D) 5

Answers

1 A

2 D

3 C

4 28

5 −3

6 C

7 −21

8 5

9 D

10 D

References

Exercise #19

Exercise #21

Exercise #22

Exercise #23

Exercise #25

Exercise #26

Exercise #23

Exercise #20

Exercise #24

Exercise #21

Notes

_____

_____

_____

_____

_____

_____

_____

_____

_____

_____

_____

_____

_____

_____

_____

Technique

#4

Finding the Perfect Relationship

Finding the Perfect Relationship

Technique #4 explores how equations relate. You'll learn to interpret visual relationships—when equations match, differ, or overlap—and what those relationships mean for their solutions.

Types of Relationships

Parallel Lines have the same slope but different y-intercepts. Because they never meet, they have no points—or solutions—in common.

Perpendicular Lines meet once at a right angle. They have slopes that are negative reciprocals (such as 2 and $-\frac{1}{2}$) and share exactly one solution.

Overlapping Graphs represent equivalent equations. Every point on one graph lies on the other, meaning they share all the same solutions.

Clear the Clutter, Find the Match

While this guide shows all graphs so you can follow each step, get in the habit of hiding extra graphs when you work on your own. Doing so makes it much easier to spot the right relationship.

BEFORE HIDING

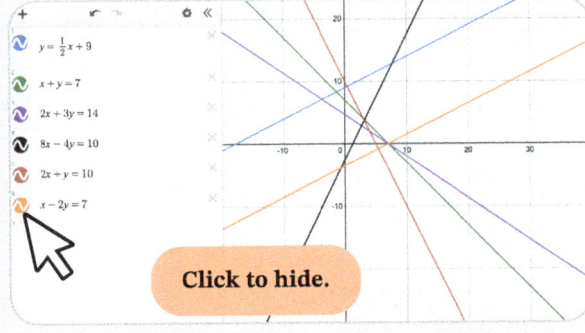

AFTER HIDING

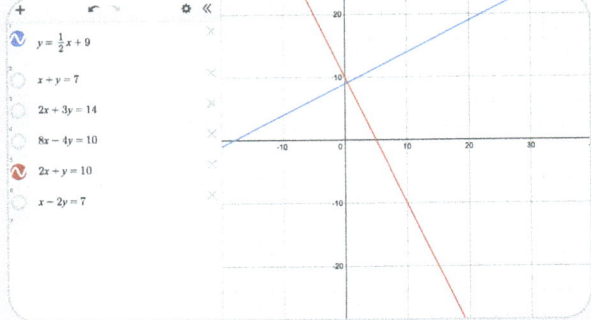

Exercise #27

$$2x + 3y = 6$$

$$6y = 12 - 4x$$

How many solutions are there to the system of equations above?

(A) Zero

(B) One

(C) Two

(D) Infinitely many

How to Solve

Step 1 Enter the given equations.

Step 2 Hide the visible graph to reveal the line underneath. Because the lines overlap, the system has infinitely many solutions.

The answer is D.

$2x + 3y = 6$

$6y = 12 - 4x$

Overlapping graphs

powered by desmos

In this guide, a **system** refers to a **system of equations** unless otherwise noted.

$y = 2x + 5$

One equation in a system is shown above. If the system has no solutions, which of the following could be the second equation?

(A) $\frac{1}{2}x + y = 5$

(B) $-2x + y = 7$

(C) $4x + y = 10$

(D) $4x + 2y = 5$

Note!

Parallel lines have no solutions in common because they never intersect.

How to Solve

Step 1 Enter the given equation and each answer choice.

Step 2 Identify the line that is parallel to the given equation.

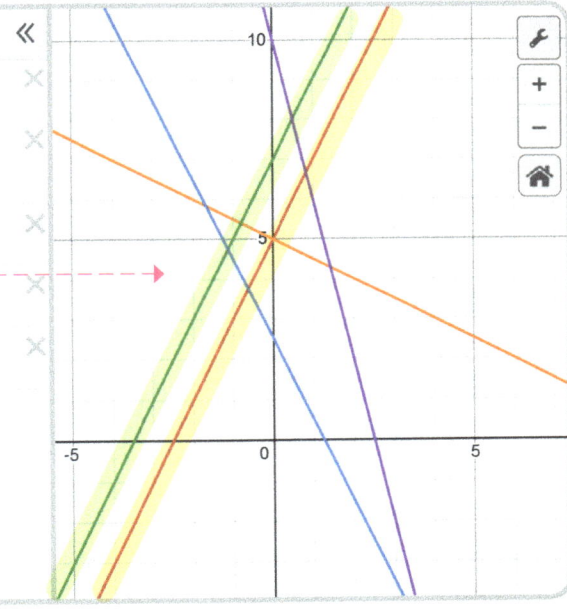

The answer is B.

Keep an algebraic method ready for cases when spotting parallel lines is tricky.

$(x^5y^3z^{-2})(xyz^{-4})$

Which of the following is equivalent to the expression above?

(A) $x^5y^{-6}z^{-8}$ (B) x^5yz^{-4}

(C) $x^6y^4z^{-6}$ (D) $x^6y^{-6}z^2$

Note!

Since Desmos does not graph expressions, convert each one into an equation by setting it equal to 100.

The expression in the question becomes

$(x^5y^3z^{-2})(xyz^{-4}) = 100$

How to Solve

Step 1 Enter the given expression and each answer choice, setting each one equal to 100. Add a slider for z.

Step 2 Identify the graph that overlaps with the first equation.

The answer is C.

$(x^5y^3z^{-2})(xyz^{-4}) = 100$

$z = 1$

-10 ———•——— 10

$x^5y^{-6}z^{-8} = 100$

$x^5yz^{-4} = 100$

$x^6y^4z^{-6} = 100$

$x^6y^{-6}z^2 = 100$

powered by desmos

If the graphs always match, so do the expressions.

If more than one graph overlaps, move the slider. Only the equivalent equation will stay aligned with the original graph.

Exercise #30

What is the equation of the line that passes through $(0, -4)$ and is parallel to $2y + 32 = 6x$?

Note!

Don't pick any parallel line! It must go through $(0, -4)$.

(A) $y + 4x = 0$

(B) $2y - 6x = 10$

(C) $3y - 9x = -12$

(D) $y + 4x = 3$

How to Solve

Step 1 Enter and label the given point.

Step 2 Enter the given equation and each answer choice.

Step 3 Find the equation whose graph is parallel to $2y + 32 = 6x$ and passes through $(0, -4)$.

The answer is C.

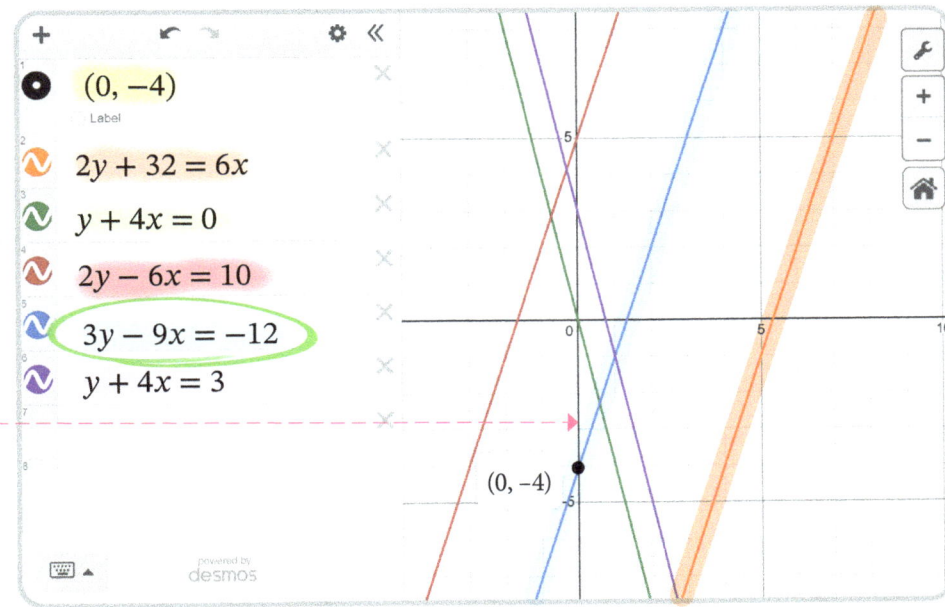

$(0, -4)$
Label

$2y + 32 = 6x$

$y + 4x = 0$

$2y - 6x = 10$

$3y - 9x = -12$

$y + 4x = 3$

$(0, -4)$

powered by
desmos

Parallel lines—forever side by side, kind of like my legs.

$$3x + 4y = 1$$

Which of the following is perpendicular to the line above?

(A) $\frac{8}{3}x + 2y = 1$

(B) $\frac{3}{8}x + 2y = 1$

(C) $-\frac{3}{8}x + 2y = 1$

(D) $-\frac{8}{3}x + 2y = 1$

How to Solve

Step 1 Enter the given equation and each answer choice.

Step 2 Identify the line that is perpendicular to the given equation. Hide graphs as needed to see the perpendicular relationship clearly.

The answer is D.

$3x + 4y = 1$

$\frac{8}{3}x + 2y = 1$

$\frac{3}{8}x + 2y = 1$

$-\frac{3}{8}x + 2y = 1$

$-\frac{8}{3}x + 2y = 1$

powered by
desmos

Psst . . . paper corners make right angles! Hold one up to your lines as I'm doing here to see if the lines look perpendicular.

Exercise #32

$$3x + 4y = 1$$
$$cx + dy = 1$$

In the system above, c and d are constants. The graphs of these two equations represent two perpendicular lines. Which of the following pairs of equations also represents two perpendicular lines?

(A) $6x + 4y = 1$
$cx - dy = 1$

(B) $6x + 4y = 1$
$cx + 2dy = 1$

(C) $6x + 4y = 1$
$2cx + dy = 1$

(D) $3x - 4y = 1$
$cx + dy = 1$

Note!

To help create perpendicular lines, use Desmoto's tip from Exercise #31.

How to Solve

Step 1 Enter the given equations. Add sliders for c and d.

Step 2 Adjust the sliders until the given equations look perpendicular.

Step 3 With the slider values unchanged, enter each answer choice to see which pair of equations creates perpendicular lines.

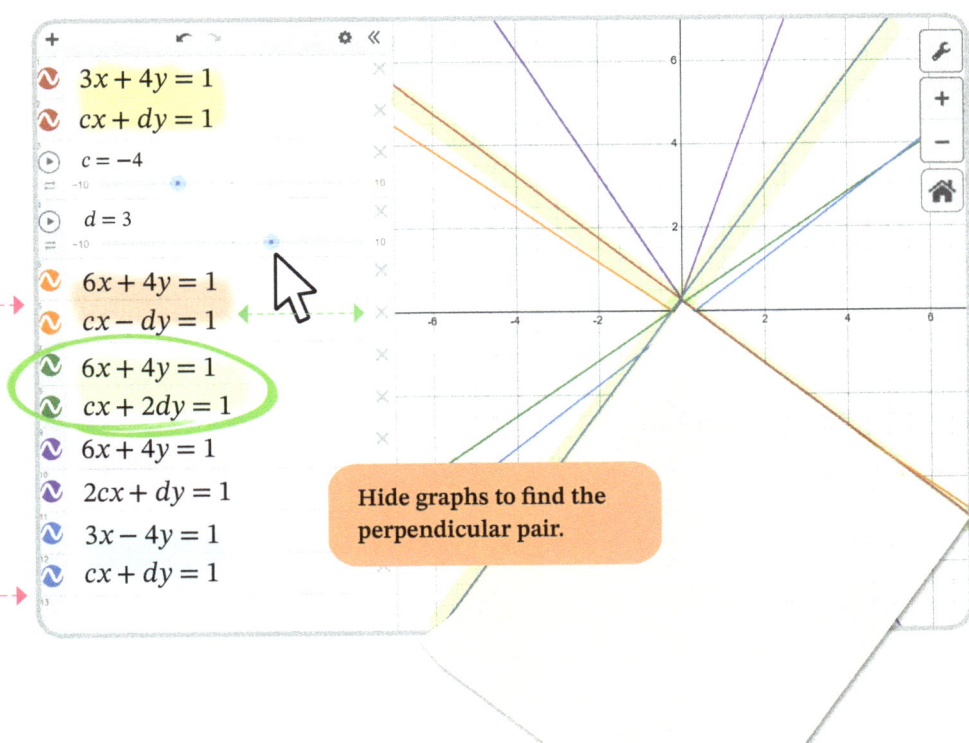

$3x + 4y = 1$
$cx + dy = 1$
$c = -4$
-10
$d = 3$
-10
$6x + 4y = 1$
$cx - dy = 1$
$6x + 4y = 1$
$cx + 2dy = 1$
$6x + 4y = 1$
$2cx + dy = 1$
$3x - 4y = 1$
$cx + dy = 1$

Hide graphs to find the perpendicular pair.

The answer is B.

Exercise #33

If $\dfrac{a^2 - b^2}{c} = 5$, which of the following must also be true?

(A) $a + b = \dfrac{5c}{a - b}$ (B) $a + b = \dfrac{a - b}{5c}$

(C) $b = \dfrac{5c}{a + b}$ (D) $c = 5(a^2 - b^2)$

Note!

To make this graphable, replace a with x and b with y: $\dfrac{x^2 - y^2}{c} = 5$

Do the same for the answer choices. Rewrite choice D: $5(x^2 - y^2) = c$

(See note below.)

How to Solve

Step 1 Enter the given equation and each answer choice, replacing a with x and b with y.

Step 2 Add a slider for c.

Step 3 Identify the graph tht overlaps with the given equation.

The answer is A.

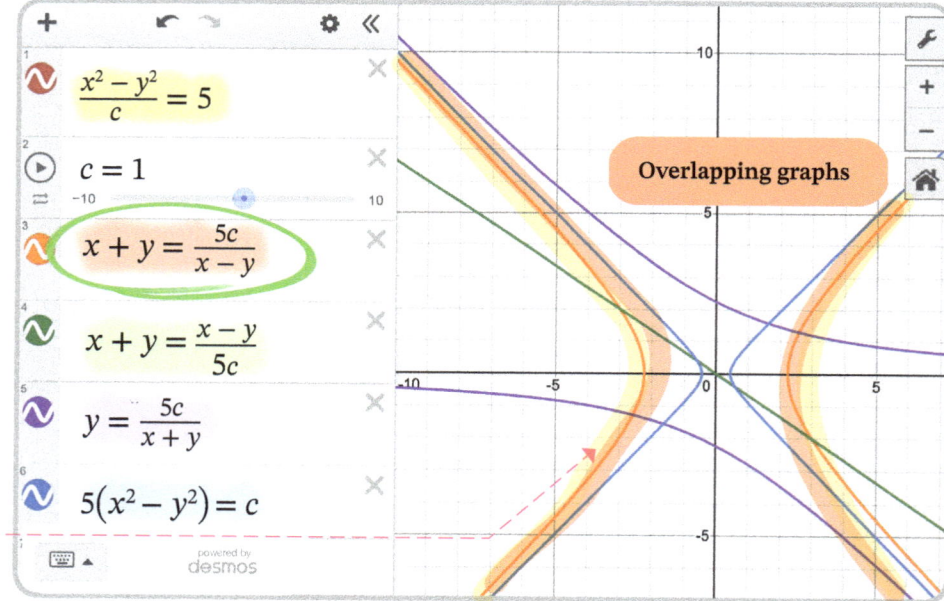

Desmos doesn't allow two equations to start with $c =$. Since the slider already uses c, write answer choice D as $5(x^2 - y^2) = c$.

Exercise #34

Which expression is equivalent to $x^{\frac{9}{10}}$ when $x > 0$?

(A) $\sqrt[10]{x^{90}}$

(B) $\sqrt[100]{x^{81}}$

(C) $\sqrt[81]{x^{100}}$

(D) $\sqrt[100]{x^{90}}$

How to Solve

Step 1 Set the expressions equal to y and enter them as shown.

Step 2 Identify the graph that overlaps with the given expression when $x > 0$.

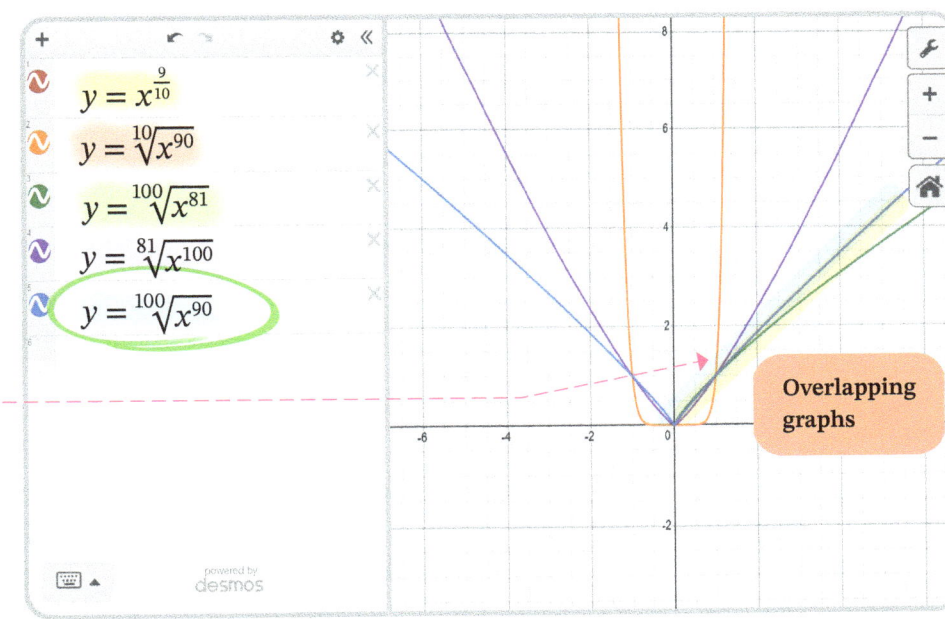

$y = x^{\frac{9}{10}}$

$y = \sqrt[10]{x^{90}}$

$y = \sqrt[100]{x^{81}}$

$y = \sqrt[81]{x^{100}}$

$y = \sqrt[100]{x^{90}}$

Overlapping graphs

The answer is D.

Overlapping graphs are like secret layers. Too bad it's not a layer cake.

Exercise #35

$$f(x) = 1.84^{\frac{x}{12}}$$

The function above can also be expressed as

$$f(x) = \left(1 + \frac{p}{100}\right)^x$$

where p is a constant. Which of the following values is closest to p?

(A) 4   (B) 5   (C) 6   (D) 7

Note!

Zoom out to see the graph better.

How to Solve

Step 1 Enter the two equations.

Step 2 Add a slider for p.

Step 3 Adjust the slider until the graphs generally overlap. Choose the answer closest to p.

The answer is B.

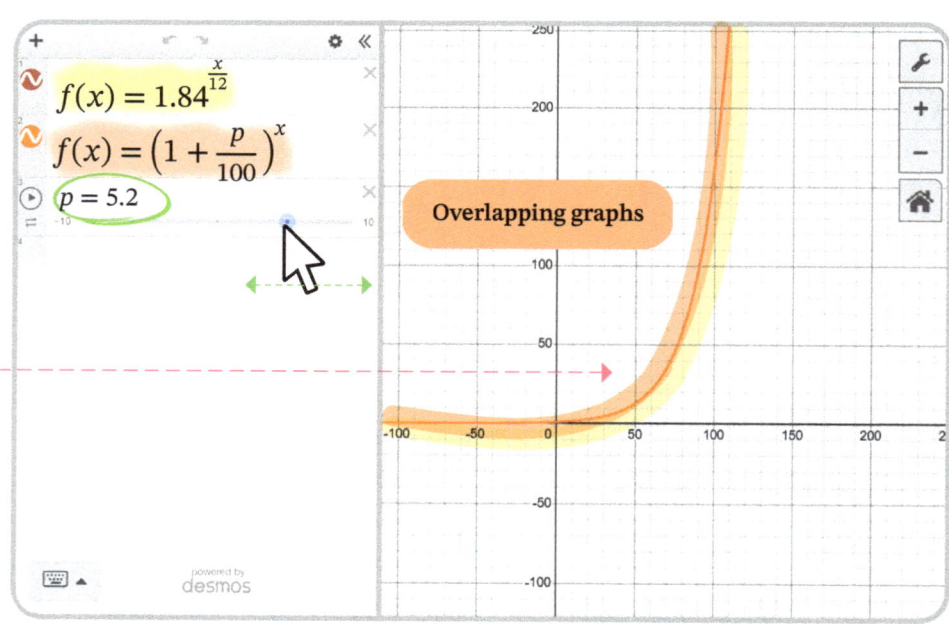

$$f(x) = 1.84^{\frac{x}{12}}$$

$$f(x) = \left(1 + \frac{p}{100}\right)^x$$

$p = 5.2$

Overlapping graphs

powered by desmos

A perfect overlap is not required. The goal is a reasonable estimate—sufficient to eliminate the answer choices.

Technique #4
PRACTICE PROBLEMS

Problem 1

$$\frac{1}{\frac{2}{x+1} + \frac{1}{x+2}}$$

If $x > -1$, which of the following is equivalent to the expression above?

A) $\frac{2x+3}{3}$

B) $\frac{x}{3} + 1$

C) $\frac{x^2 + 3x + 2}{3x + 5}$

D) $\frac{2x+5}{x^2 + 3x + 2}$

Problem 2

What is the equation of the line that passes through $(9, 4)$ and is parallel to $6x = 10y + 20$?

A) $5x + 3y = 12$

B) $5x - 3y = 6$

C) $3x + 5y = 10$

D) $3x - 5y = 7$

Problem 3

$$\frac{a^4 b^2 c^{-3}}{a^2 b^{-1} c^2}$$

Which expression is equivalent to the expression above, where a, b, and c are positive?

A) $a^2 b^3 c^{-5}$
B) $a^2 b^3 c^{-1}$
C) $a^2 b^{-3} c^{-1}$
D) $a^2 b^{-1} c^{-5}$

Hint: Choices A and B will both overlap the given expression when $c = 1$. As you move the slider for c, only the correct answer will maintain the overlap.

Problem 4

$$3x - 4y = -4$$

What is the equation of a line that has the same y-intercept and is perpendicular to the line above?

A) $y = 3x + 4$

B) $y = \frac{1}{3}x - 4$

C) $y = -\frac{4}{3}x + 4$

D) $y = -\frac{4}{3}x + 1$

Technique #4
PRACTICE PROBLEMS

Problem 5

The function f is defined by $f(x) = 2.03^{\frac{x}{10}}$. If $f(x)$ can be written in the form

$$f(x) = \left(1 + \frac{p}{100}\right)^x$$

for some constant p, which of the following is closest to p?

A) 7.3
B) 7.6
C) 7.9
D) 8.2

Problem 6

$$x = 7 - 2y$$
$$y = (x + 1)(3 - 2x)$$

How many solutions are there to the system of equations above?

A) Zero
B) One
C) Two
D) Infinitely many

Problem 7

$$5x - 2y = 7$$
$$mx + ny = 1$$

The system above consists of two perpendicular lines, where m and n are constants. Which of the following pairs of equations also represents two perpendicular lines?

A) $5x - 2y = 7$
$mx - 2ny = 1$

B) $5x - 2y = 7$
$mx + 2ny = 1$

C) $5x + 2y = 7$
$mx - ny = 10$

D) $5x + 2y = 7$
$2mx + 2ny = 1$

Technique #4
PRACTICE PROBLEMS

Problem 8

Which expression is equivalent to $a^{\frac{5}{6}}$ when $a > 0$?

A) $\sqrt[50]{a^{60}}$

B) $\sqrt[30]{a^{25}}$

C) $\sqrt[36]{a^{25}}$

D) $\sqrt[15]{a^{18}}$

Problem 9

$$\frac{2x - 4y}{\frac{2}{3}} = 12a$$

Which equation correctly expresses y in terms of x and a?

A) $y = \frac{7x + 8a}{4}$

B) $y = \frac{x - 4a}{2}$

C) $y = -\frac{7x - 8a}{4}$

D) $y = 7x + 8a$

Problem 10

$$\frac{8}{5}y - \frac{2}{3}x = \frac{7}{20}$$
$$12y = 25px + 10p$$

For the system of equations above, for what value of p are there no solutions?

A) $\frac{1}{5}$

B) $\frac{1}{2}$

C) $\frac{3}{5}$

D) $\frac{5}{4}$

Answers

1 C

2 D

3 A

4 D

5 A

6 A

7 C

8 B

9 B

10 A

References

Exercise #34

Exercise #30

Exercise #29

Exercise #31

Exercise #35

Exercise #27

Exercise #32

Exercise #34

Exercise #33

Exercise #28

Notes

_____

_____

_____

_____

_____

_____

_____

_____

_____

_____

_____

_____

_____

_____

Technique

5

The Hidden
Equation

The Hidden Equation

Technique #5 introduces regression—a method for finding an equation that fits a set of points. Desmos analyzes the numerical pattern in those points and generates an equation that models them, then graphs the result so you can see how the equation behaves. This technique includes two approaches: the General Method, used when enough points are given, and the Custom Method, used when fewer points are provided along with extra information. In both cases, you'll use Table ⊞ to enter the coordinate points.

The General Method

The General Method uses Desmos to generate an equation that best fits the given points. Enter the coordinate points into a table—this unlocks regression features. Once at least two points are entered, **Add Regression** 🔧 will appear to the left of the table. Select it, then choose the appropriate regression model (linear, quadratic, etc.) from the dropdown menu. Desmos automatically calculates the best-fit equation and graphs it. To use the General Method, at least two points are needed for a linear equation (line) and three points for a quadratic (parabola). With fewer points, the model will not be accurate.

Example

Let's say a question gives you the points $(-2, 15)$, $(1, 3)$, and $(6, 23)$ and asks you to find the quadratic equation that passes through them. Here's how to do it.

Step 1 Click **Add Item** ⊕ and select **Table.** Enter $(-2, 15)$, $(1, 3)$, and $(6, 23)$.

Step 2 Click **Add Regression** 🔧.

Step 3 Select **Quadratic Regression** from the dropdown menu. Desmos will instantly calculate the quadratic equation and graph it.

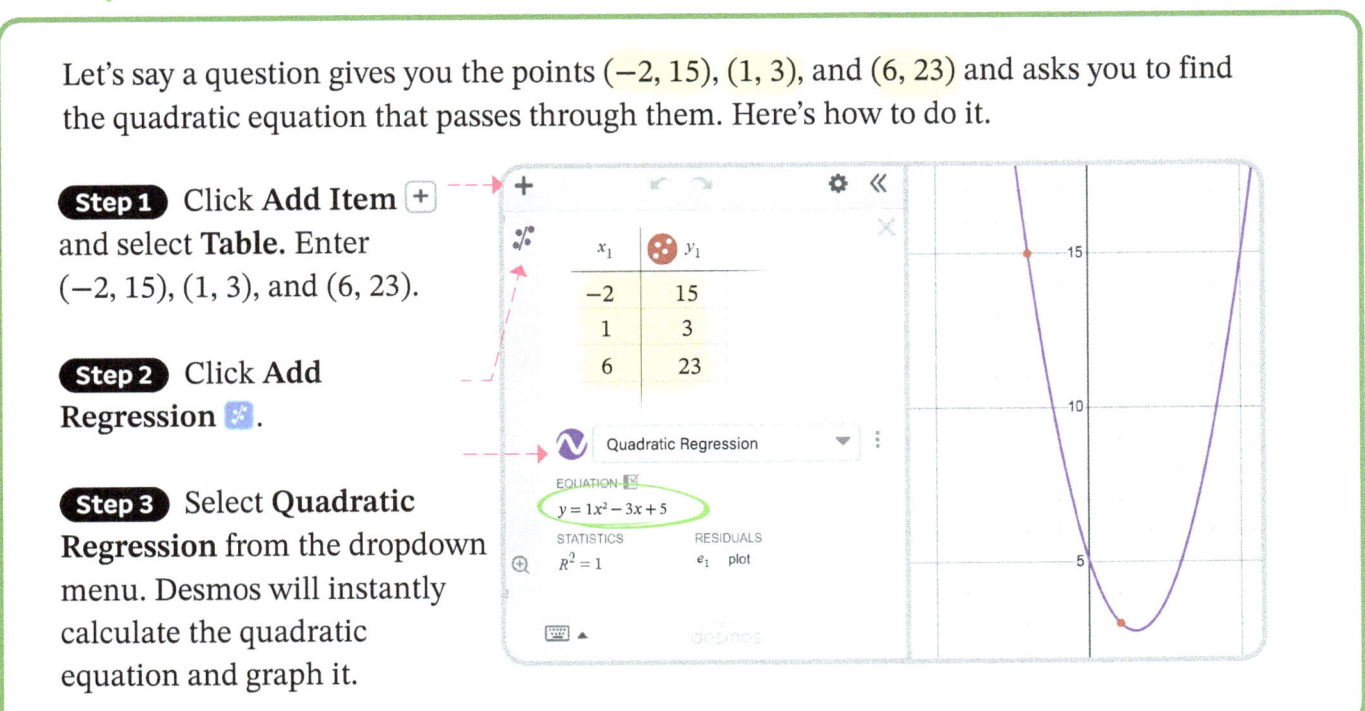

The Custom Method

Custom regressions help you find an equation when you're given fewer points but part of the equation is already known. Instead of choosing a model from the dropdown menu, you'll enter the points and type a regression model that includes the given information, such as the slope of a line. This method requires more careful thinking than the General Method, but it allows you to solve a wider range of SAT questions.

Regression Type	Regression Model
Linear (Slope-Intercept Form)	$y_1 \sim mx_1 + b$, where m is the slope and b is the y-intercept
Quadratic (Standard Form)	$y_1 \sim ax_1^2 + bx_1 + c$
Quadratic (Vertex Form)	$y_1 \sim a(x_1 - h)^2 + k$ where (h, k) is the vertex

Example

If a question provides the point (6, 13) and the slope of a line ($m = 5$), you can use a custom regression to find the equation of the line.

Step 1 Click **Add Item** ➕ and select **Table** 🧮. Enter (6, 13).

Step 2 In the next row, enter the linear regression model, replacing m with 5. Desmos will graph the line and show the coefficient b. The equation is $y = 5x - 17$.

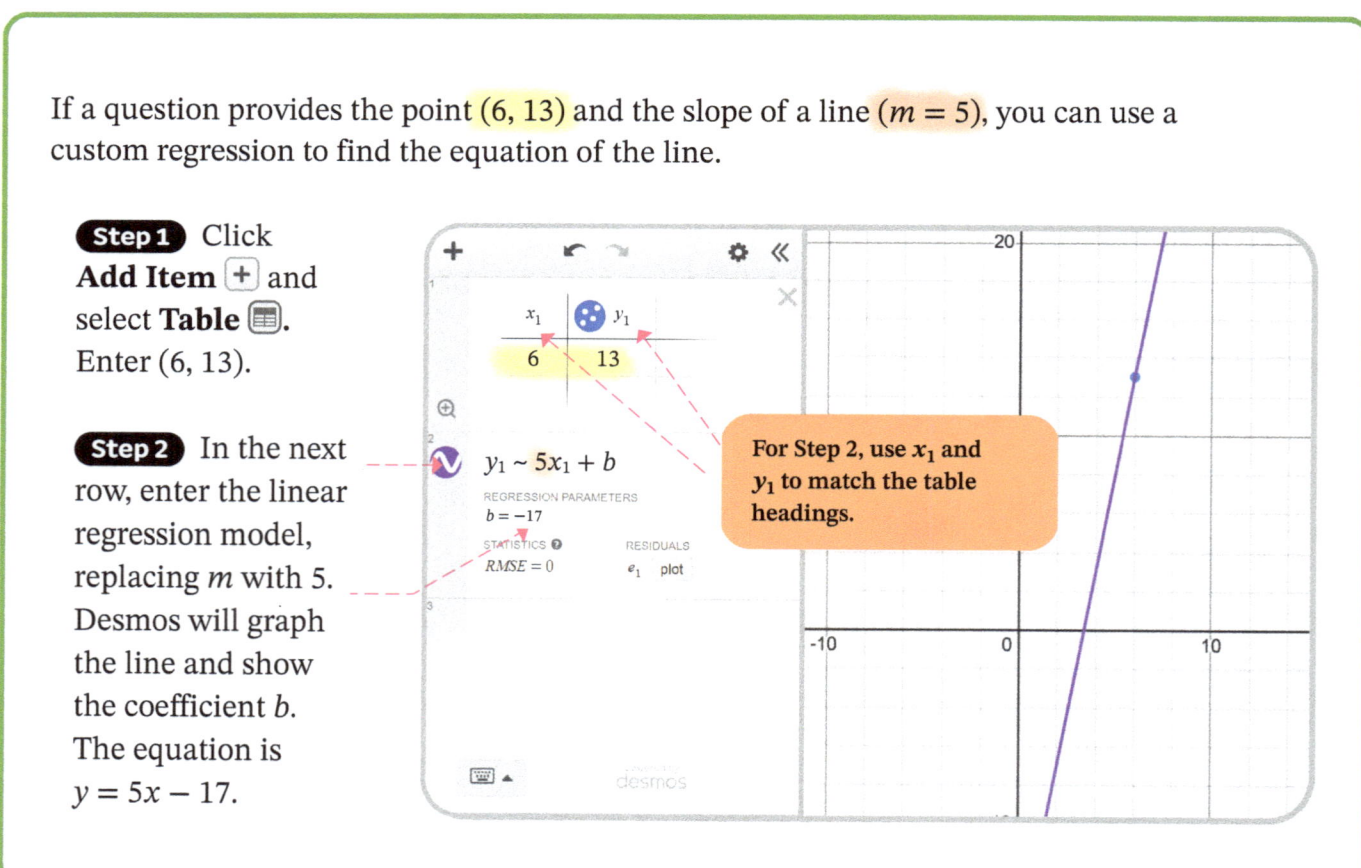

For Step 2, use x_1 and y_1 to match the table headings.

$y_1 \sim 5x_1 + b$

REGRESSION PARAMETERS
$b = -17$

STATISTICS ❓   RESIDUALS
$RMSE = 0$ e_1 plot

Typing Basics for Regression Models

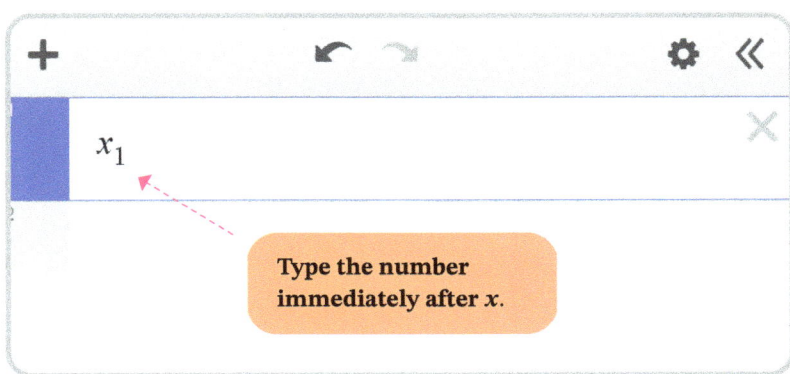

The Tilde (~): Use a tilde (~) instead of an equal sign when typing regression models. On the Desmos Keypad, tap **ABC**, then select the tilde (~). On a computer keyboard, the tilde is located near the top left,

Subscript Notation (x_1 and y_1): The variables x and y must use the same subscripts as the table headings where you enter your coordinate points. For example, if the table headings are x_1 and y_1, use x_1 and y_1 in your regression model. To enter a subscript, type the number immediately after x or y, and Desmos will format it as a subscript.

x_1

Type the number immediately after x.

Typing Basics for the Custom Method

Math Notation	Desmos Keypad	Computer or Tablet Keyboard
Tilde (~)	» Click ABC » Click ~	» Type ~
Subscript e.g., x_1	» Click x » Click 1 » (Desmos will automatically make the 1 a subscript.)	» Type x » Type 1 » (Desmos will automatically make the 1 a subscript.)

Exercise #36

The points (2, 5) and (6, 13) are on a line. If the line is written in the form $y = mx + b$, what is the value of $m + b$?

(A) 3   (B) 4   (C) 5   (D) 6

How to Solve

Regression symbol appears here.

Step 1 Click **Add Item** ⊕ and select **Table** ▦. Enter the given points.

Step 2 Click **Add Regression** and select **Linear Regression** from the dropdown menu.

Step 3 Determine the m and b values from the given equation and add the values to find their sum.

The answer is A.

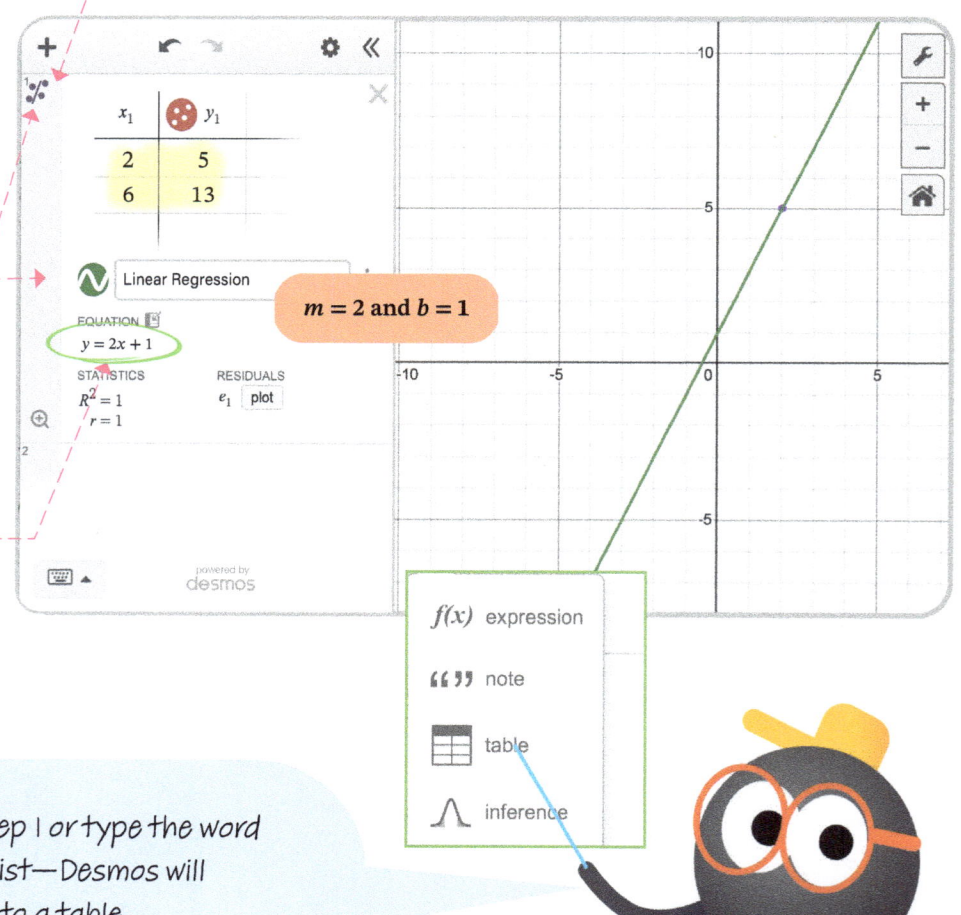

To create a table, follow Step 1 or type the word **table** into the Expression List—Desmos will automatically convert it into a table.

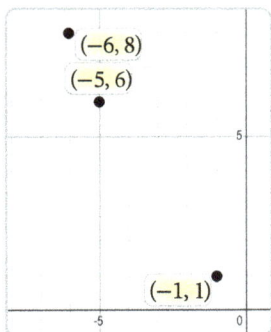

Note!

A regression finds a line that follows the overall pattern of the points in a scatter plot.

Which of the following equations best represents a linear model for the data in the scatter plot above?

(A) $y = -x + 3$

(B) $y = -0.5x + 2$

(C) $y = -1.3x - 0.5$

(D) $y = -4x + 2$

How to Solve

Step 1 Click **Add Item** ⊞ and select **Table** 🗏. Enter the given points.

Step 2 Click **Add Regression** 🔣 and select **Linear Regression**.

Step 3 Compare the regression equation to the answer choices and pick the closest match.

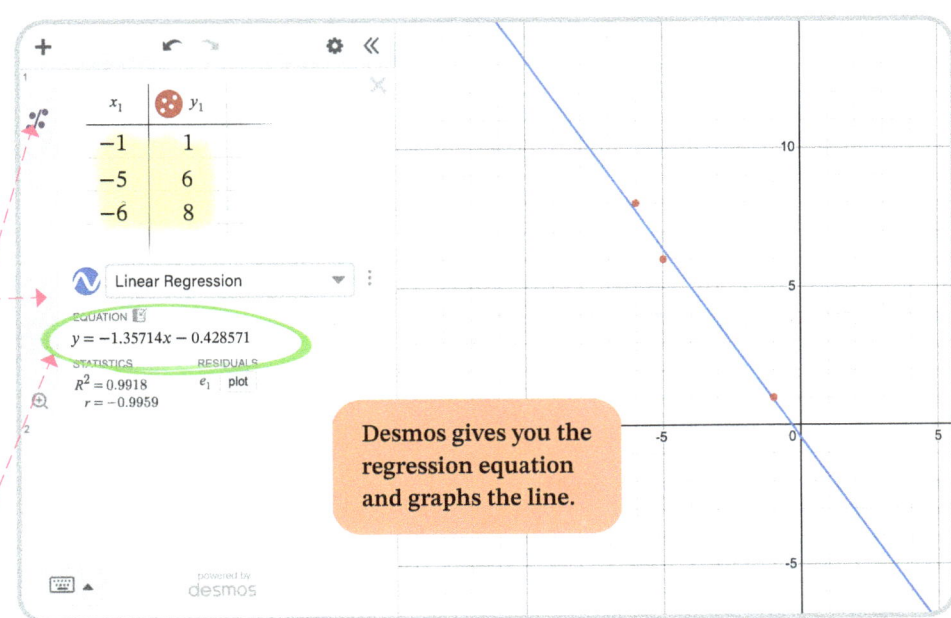

Desmos gives you the regression equation and graphs the line.

The answer is C.

Note!

To find $f(4)$, identify the y value when $x = 4$.

The points $(-2, -2.5)$, $(1, 8)$, and $(6, 25.5)$ lie on graph of the linear function f. What is the value of $f(4)$?

(A) $15\frac{1}{3}$ (B) $15\frac{1}{4}$ (C) $16\frac{1}{2}$ (D) $18\frac{1}{2}$

How to Solve

Step 1 Click **Add Item** ⊕ and select **Table** 🗐. Enter the given points.

Step 2 Click **Add Regression** 🗲 and select **Linear Regression**.

Step 3 Click and drag your mouse along the line to find the y value when $x = 4$.

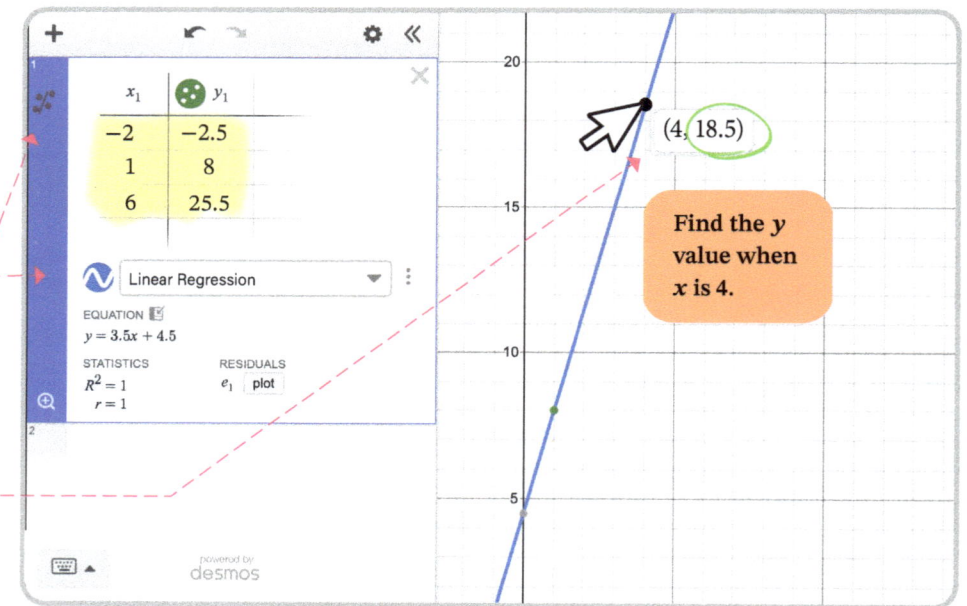

Find the y value when x is 4.

The answer is D.

For Step 3, you can also enter $x = 4$ into a new row and determine the y value of the intersection.

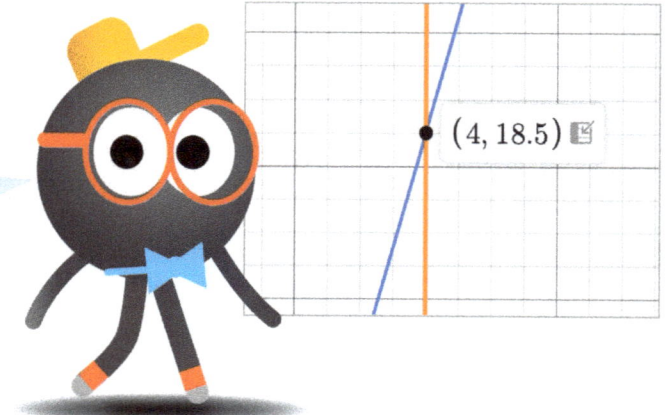

Exercise #39

Line p passes through $\left(\frac{1}{2}, \frac{3}{4}\right)$ and $(-1, d)$ and has a slope of -5. What is the value of d?

Note!

Use the Custom Method here because you're given one point and the slope.

Linear regression model:
$y_1 \sim mx_1 + b$
Replace m with -5:
$y_1 \sim -5x_1 + b$

How to Solve

Step 1 Click **Add Item** ⊕ and select **Table** ▦. Enter $\left(\frac{1}{2}, \frac{3}{4}\right)$.

Step 2 Input the linear regression model, replacing m with -5. Make sure to use the tilde ($\sim$) and x_1 and y_1.

Step 3 Click and drag along the line to find the value of y when $x = -1$.

The answer is 8.25 or $\frac{33}{4}$.

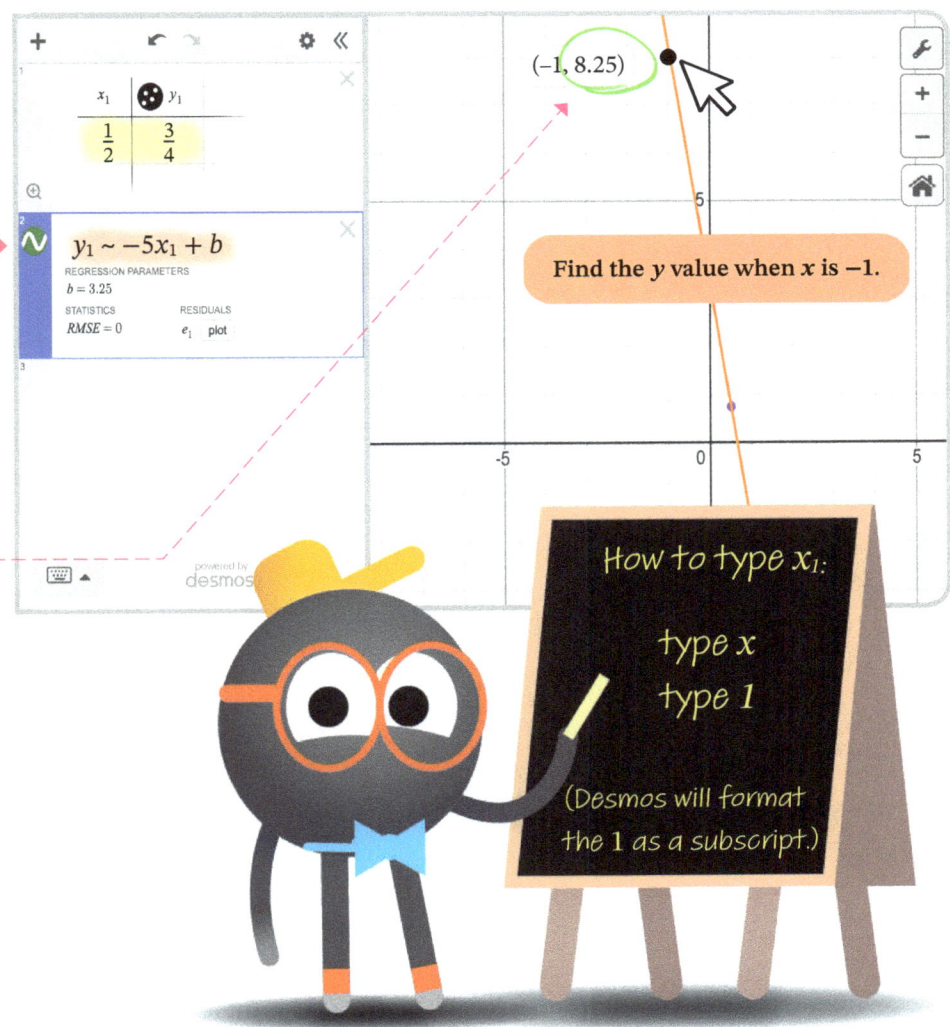

(−1, 8.25)

Find the y value when x is −1.

How to type x_1:

type x
type 1

(Desmos will format the 1 as a subscript.)

Exercise #40

Line *n* passes through (−12, 4) and is perpendicular to a line with a slope of 4. What is the *x*-intercept of line *n*?

(A) (4, 0)　　　　　(B) (5, 0)

(C) (6, 0)　　　　　(D) (7, 0)

Note!

Use the Custom Method here because you're given one point and you can determine the slope.

Linear regression model:
$$y_1 \sim mx_1 + b$$
Replace *m* with $-\frac{1}{4}$:
$$y_1 \sim -\frac{1}{4}x_1 + b$$

How to Solve

Step 1 Click **Add Item** ⊕ and select **Table** 🗐. Enter the given point.

Step 2 Input the linear regression model with $m = -\frac{1}{4}$.

Step 3 Click the *x*-intercept to find the answer.

The answer is A.

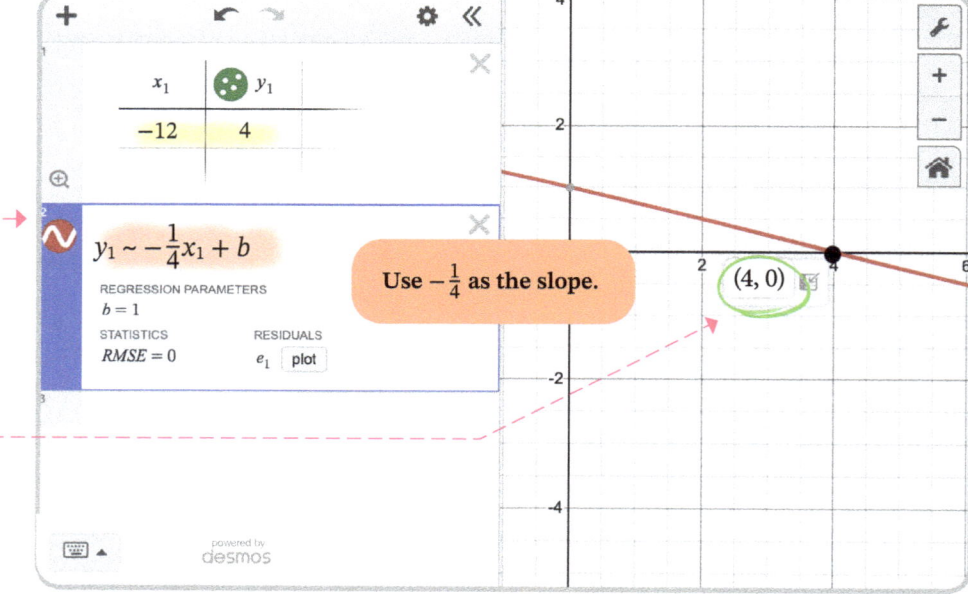

Use $-\frac{1}{4}$ as the slope.

Remember that slopes of perpendicular lines are negative reciprocals of each other.

m　　$-\dfrac{1}{m}$

Line *l* has the equation $4x + 2y = 40$. Line *m* is formed by translating line *l* six units to the left. If the *y*-intercept of line *m* is $(0, n)$, what is the value of *n*?

_____

Note!

This visual method can be especially helpful if you haven't yet learned how to perform translations algebraically.

How to Solve

Step 1 Enter the equation for line *l* and shift the intercepts 6 units to the left to identify points on line *m*: $(4, 0)$ and $(-6, 20)$.

Step 2 Click **Add Item** and select **Table** 🗐. Enter $(4, 0)$ and $(-6, 20)$.

Step 3 Click **Add Regression** and select **Linear Regression**.

Step 4 Click the *y*-intercept of line *m* to find the *y* value.

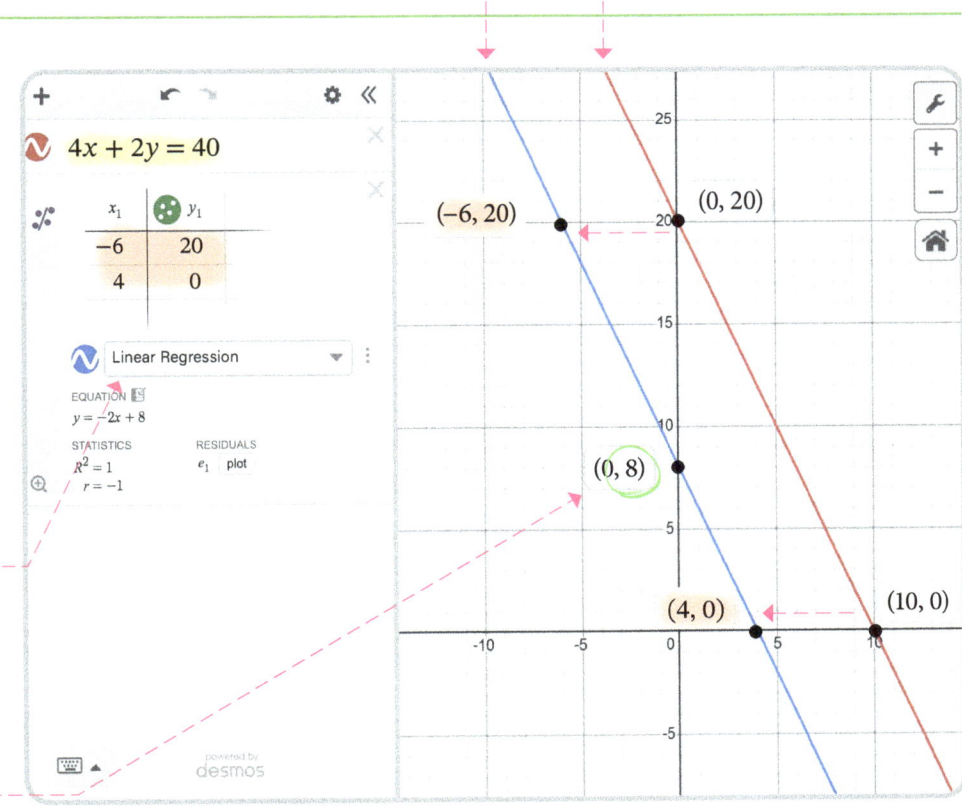

The answer is 8.

when translating points, take a quick look to make sure you entered them correctly. A visual check helps catch mistakes.

Note!

with three exact points on a parabola, a quadratic regression gives the exact equation.

A parabola passes through $(-2, 1)$, $(-1, 2)$, and $(0, 5)$. If the quadratic equation is written in the form $y = ax^2 + bx + c$, what is the value of b?

(A) 1   (B) 2   (C) 3   (D) 4

How to Solve

Step 1 Click **Add Item** ⊞ and select **Table** 🗐. Enter the given points.

Step 2 Click **Add Regression** 🎛 and select **Quadratic Regression**.

Step 3 Identify the b value in the given quadratic equation.

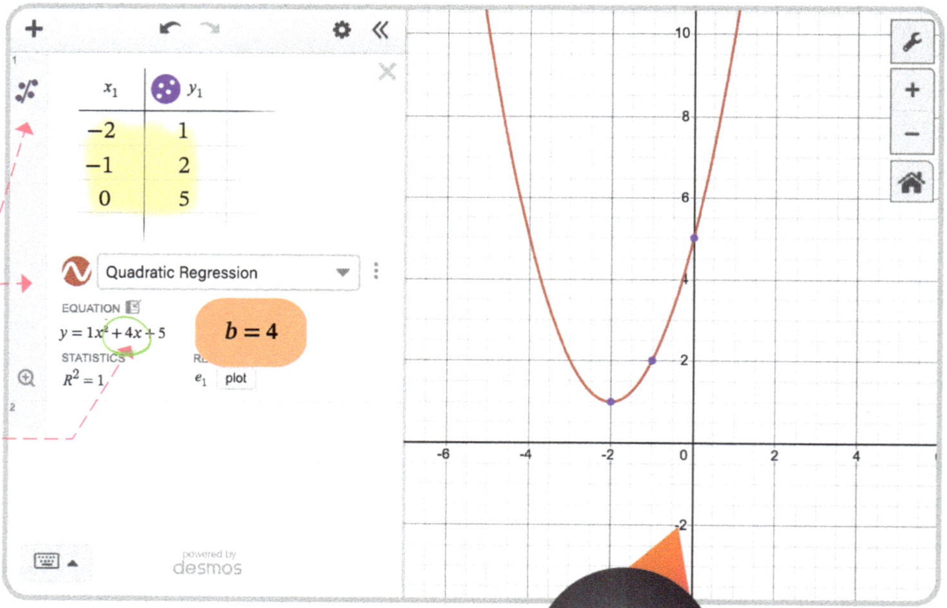

The answer is D.

x_1	y_1
-2	1
-1	2
0	5

Quadratic Regression

EQUATION
$y = 1x^2 + 4x + 5$

$b = 4$

STATISTICS
$R^2 = 1$

e_1 plot

powered by
desmos

I always give rock-solid advice. Get it? I'm literally on a rock.

Exercise #43

A parabola with equation $y = ax^2 + bx - 20$ passes through $(4, -8)$ and $(-1, -13)$. What is the sum of a and b?

(A) −5   (B) −3   (C) 2   (D) 4

Note!

Use the Custom Method because you're given two points and the constant c.

Quadratic regression model (standard form):
$$y_1 \sim ax_1^2 + bx_1 + c$$
Replace c with −20:
$$y_1 \sim ax_1^2 + bx_1 - 20$$

How to Solve

Step 1 Click **Add Item** ⊕ and select **Table** ▦. Enter the given points.

Step 2 Input the quadratic regression model in standard form, replacing c with −20.

Step 3 Determine the a and b values and add them: $2 + (-5) = -3$.

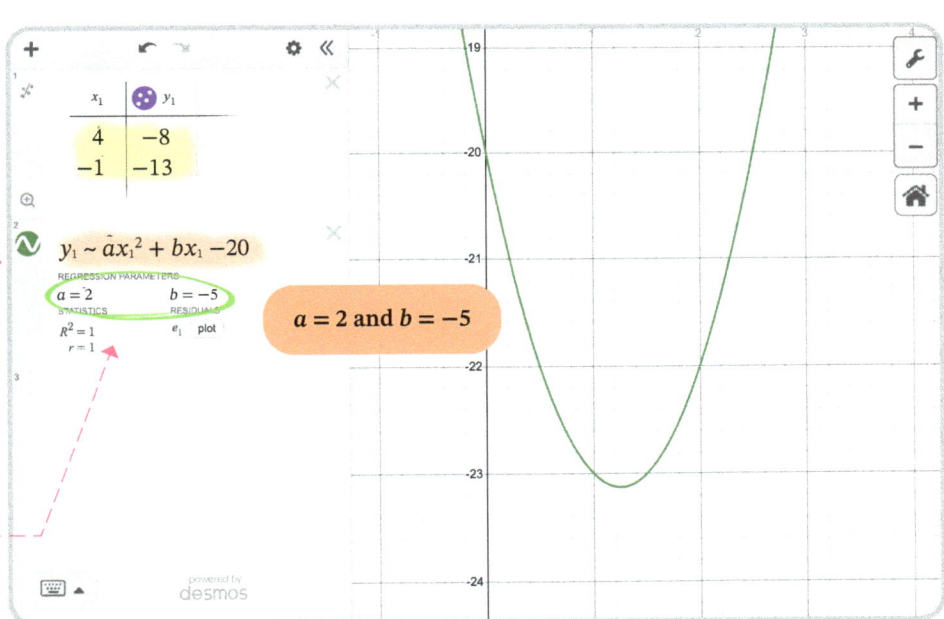

$a = 2$ and $b = -5$

The answer is B.

$$y_1 \sim ax_1^2 + bx_1 - 20$$

Did you type the regression model correctly? Sometimes I forget to write all the subscripts.

Exercise #44

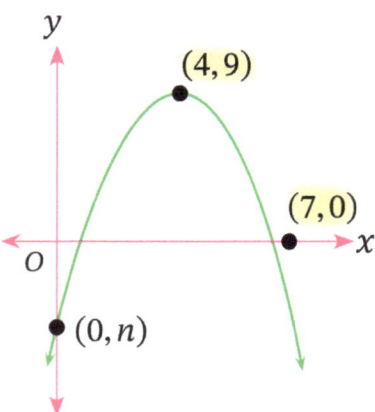

Note: Figure not drawn to scale.

The graph of the quadratic function above has a vertex of (4, 9). What is the value of n?

Note!

The vertex goes into the custom regression model and also counts as a point.

Quadratic regression model (vertex form):

$$y_1 \sim a(x_1 - h)^2 + k$$

Replace h with 4 and k with 9:

$$y_1 \sim a(x_1 - 4)^2 + 9$$

How to Solve

Step 1 Click **Add Item** + and select **Table** 📋. Enter the given points.

Step 2 Input the quadratic regression model in vertex form, replacing h with 4 and k with 9.

Step 3 Click the y-intercept of the parabola to find the value of y.

The answer is –7.

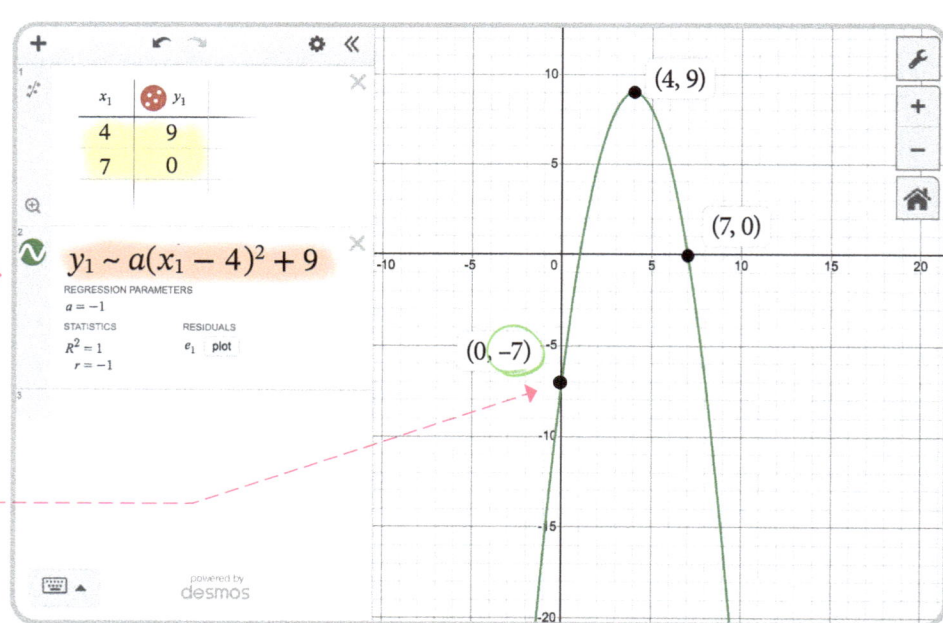

Exercise #45

A line goes through $(-5, 0)$ and $(0, -5)$. A parabola goes through $(-5, 0)$, $(-3, -4)$, and $(1, 12)$. If the two graphs intersect at (a, b), what could be the value of a?

(A) $-\dfrac{5}{2}$ (B) -2 (C) $-\dfrac{3}{2}$ (D) -3

Note!

Linear points:
$(-5, 0), (0, -5)$

Quadratic points:
$(-5, 0), (-3 , -4), (1, 12)$

How to Solve

Step 1 Click **Add Item** ⊞ and select **Table** ▦. Enter $(-5, 0)$ and $(0, -5)$.

Step 2 Click **Add Regression** ⚡ and select **Linear Regression**.

Step 3 Create **Add Item** ⊞. Select **Table** ▦. Enter $(-5, 0)$, $(-3, -4)$, and $(1, 12)$.

Step 4 Click **Add Regression** ⚡ and select **Quadratic Regression**.

Step 5 Click the intersection of the resulting graphs to find the x value.

The answer is B.

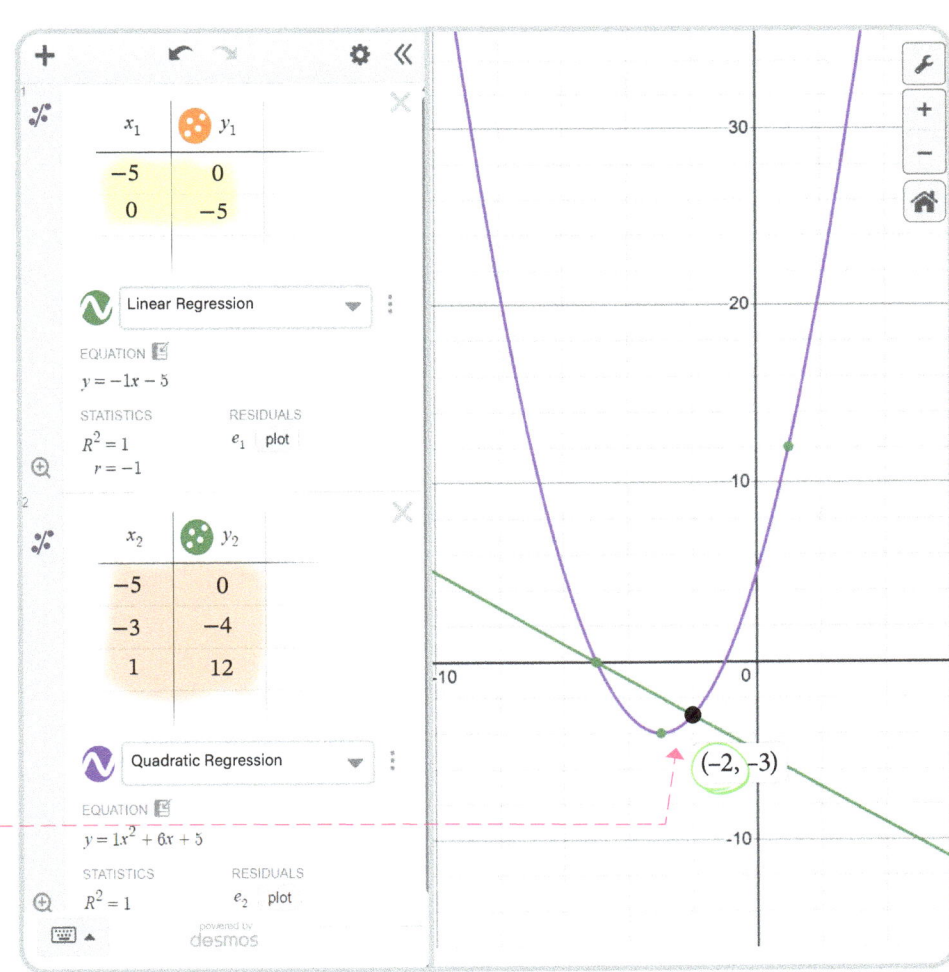

Technique #5
PRACTICE PROBLEMS

Problem 1

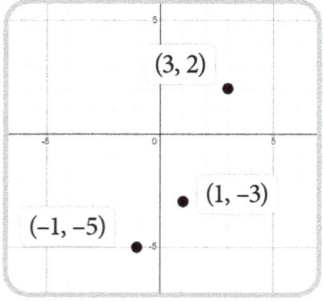

Which of the following equations best approximates a linear model for the data shown in the scatter plot above?

A) $y = 3.5x - 9.5$

B) $y = 3.9x - 3.2$

C) $y = 1.8x - 3.7$

D) $y = 3.75x - 0.8$

Problem 2

Line k passes through $(1, 4)$ and $(3, c)$ and has a slope of 2.125. What is the value of c?

Hint: Don't put $(3, c)$ in the table.

Problem 3

A parabola with vertex $(2, 4)$ passes through $(5, 10)$ and $(0, d)$. What is the value of d?

A) $\frac{20}{3}$

B) $\frac{3}{20}$

C) $-\frac{20}{3}$

D) $-\frac{3}{20}$

Problem 4

The graph of function h is a line. If $h(1) = 1$ and $h(3) = -3$, what is the value of $h(-1)$

A) -1
B) 0
C) 3
D) 5

Problem 5

x	y
-2	-12
-1	-9

The table above shows points for the linear function f. The function $g(x)$ is obtained by translating $f(x)$ up 4 units. What is the x-coordinate of the x-intercept of $g(x)$?

A) $\frac{1}{3}$

B) $\frac{2}{3}$

C) 1

D) $\frac{4}{3}$

Problem 6

A parabola passes through $(0, 4)$, $(-2, 4)$, and $(2, 20)$. If the parabola is written in the form $y = ax^2 + bx + c$, what is the value of b?

A) -2
B) 3
C) 4
D) 6

Problem 7

A line passes through $(-2, 2)$ and $(0, 6)$. A parabola passes through $(3, 17)$, $(-2, 22)$, and $(0, 8)$. What is one point of intersection between the line and the parabola?

A) $(2, 10)$
B) $(3, 11)$
C) $(2, 9)$
D) $(1, 7)$

Technique #5
PRACTICE PROBLEMS

Problem 8

Line q passes through $(4, 7)$ and is perpendicular to line with a slope of -2. What is the x-intercept of line q?

A) $(7.5, 0)$
B) $(-10, 0)$
C) $(18, 0)$
D) $(5, 0)$

Problem 9

The graph of a quadratic equation passes through $(-1, 27)$ and $(-13, -33)$ and can be written as $y = -2x^2 + bx + c$. What is the value of b?

A) -6
B) -23
C) 6
D) 23

Problem 10

x	y
1	4.5
2	6
3	7.5
4	9

If the equation $y = mx + b$ represents the data above, what is the sum of $x + b$?

_____

Hint: There are lots of points here—but remember, you only need to enter two points into the table to create a linear regression.

Technique #5
PRACTICE PROBLEMS

Answers

1 C

2 8.25 or $\frac{33}{4}$

3 A

4 D

5 B

6 C

7 A

8 B

9 B

10 4.5 or $\frac{9}{2}$

References

Exercise #37

Exercise #39

Exercise #44

Exercise #38

Exercise #41

Exercise #42

Exercise #45

Exercise #40

Exercise #43

Exercise #36

Notes

Technique

#6

When Two Equations Are Better Than One

When Two Equations Are Better Than One

Technique #6 shows how to split one equation into two, creating a system of equations. This approach is especially useful when an equation includes an unknown constant or when you need to determine how many solutions a situation has.

By graphing each side of the equation separately, you can see whether the graphs intersect, overlap, or never meet—making patterns and solutions easier to understand.

How to Convert One Equation Into a System

Create a system of equations by setting each side of the equation equal to y. Enter the system into Desmos. The solutions are the points of intersection.

Splitting One Equation Into Two

$$2x + 5c = 3x - 6$$

$$2x + 5c = y$$
$$3x - 6 = y$$

System of equations

Whether you write equations in the form $y =$ or $= y$ (for example, $y = 2x + 5c$ or $2x + 5c = y$), Desmos treats them the same way—so don't worry about which side the y is on for this technique.

Sometimes two things are better than one—two socks, two bow ties . . . and yes, two equations.

Exercise #46

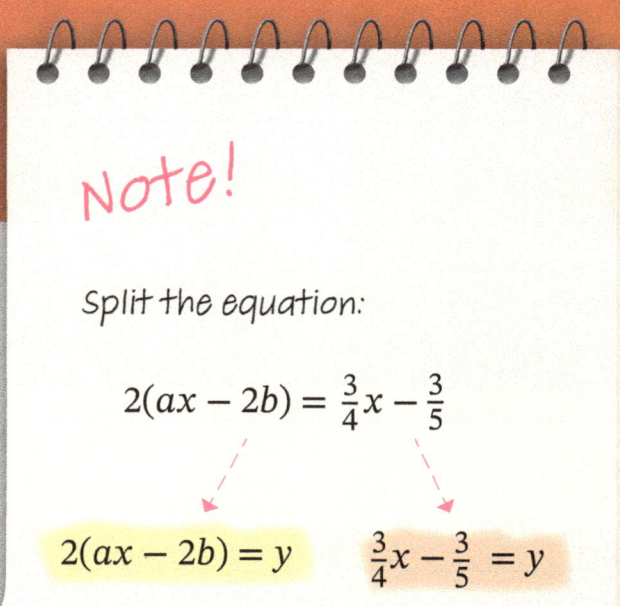

$2(ax - 2b) = \frac{3}{4}x - \frac{3}{5}$

In the equation above, a and b are constants and $b < 0$. If the equation has no solutions, what is the value of a?

(A) 0.25　(B) 0.375　(C) 0.5　(D) 0.75

Note!

Split the equation:

$2(ax - 2b) = \frac{3}{4}x - \frac{3}{5}$

$2(ax - 2b) = y$　　$\frac{3}{4}x - \frac{3}{5} = y$

How to Solve

Step 1 Enter the two split equations and add sliders for a and b.

Step 2 Set b to any value less than 0. Then test each answer choice for a—either by typing the value or using the slider—and see which value makes the two lines parallel.

The answer is B.

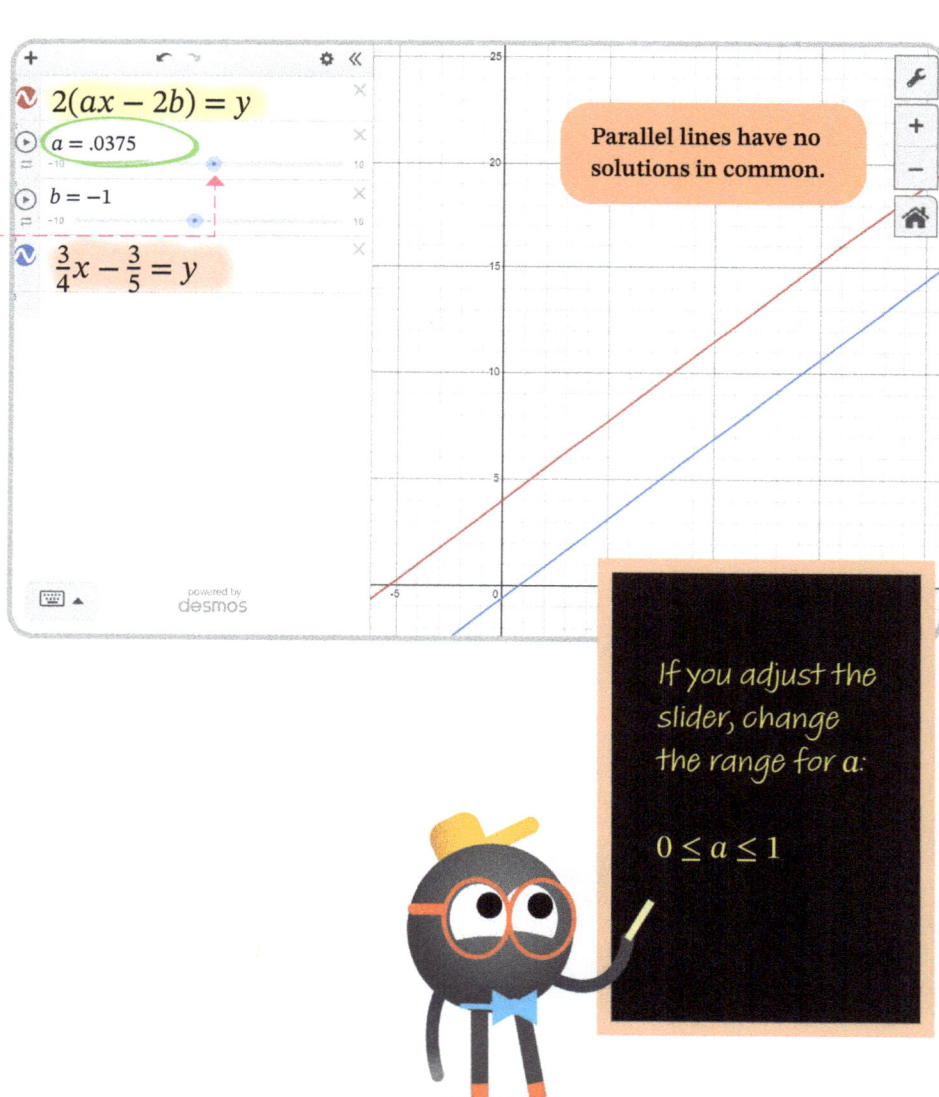

Parallel lines have no solutions in common.

If you adjust the slider, change the range for a:

$0 \leq a \leq 1$

Exercise #47

$x^2 + 3x + 7 = 0$

How many real solutions are there to the given equation?

(A) Zero

(B) One

(C) Two

(D) Infinitely many

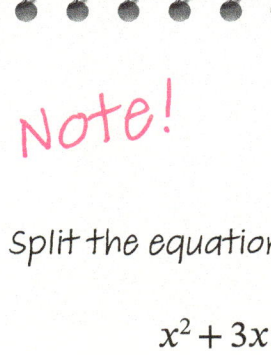

Note!

Split the equation:

$$x^2 + 3x + 7 = 0$$

$x^2 + 3x + 7 = y$ $y = 0$

How to Solve

Step 1 Enter the two split equations.

Step 2 Determine how many times the parabola intersects the line. Since there are no intersections, there are no solutions.

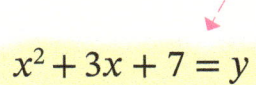

$x^2 + 3x + 7 = y$

$y = 0$

No intersections = No solutions

The answer is A.

Desmos automatically adds a slider when you type $y = 0$, but do not adjust it.

Exercise #48

The function g is defined by $g(x) = x(x - 3)(x + 5)^2$. If $g(6 - x) = 0$, what is the sum of all possible values of x?

_____

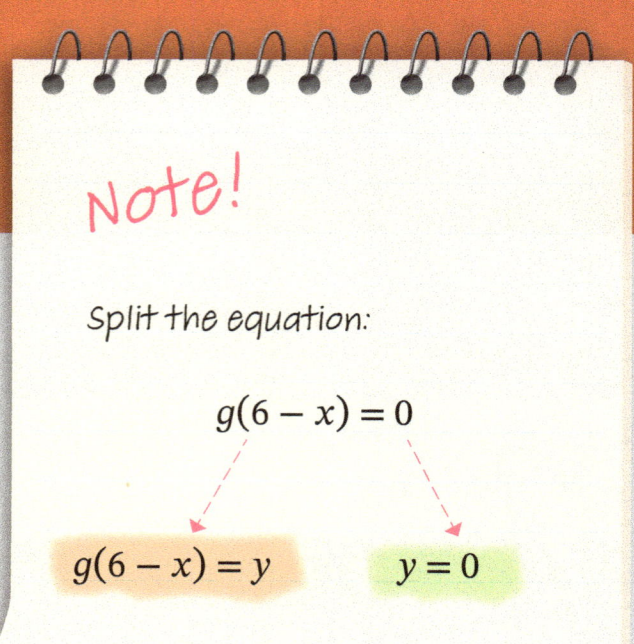

Note!

Split the equation:

$$g(6 - x) = 0$$

$$g(6 - x) = y \qquad y = 0$$

How to Solve

Step 1 Enter the function $g(x) = x(x - 3)(x + 5)^2$.

Step 2 Enter the two equations created by splitting $g(6 - x) = 0$.

Step 3 Identify the x values where $g(6 - x) = y$ intersects $y = 0$. Add the x values of the intersections: $3 + 6 + 11 = 20$.

The answer is 20.

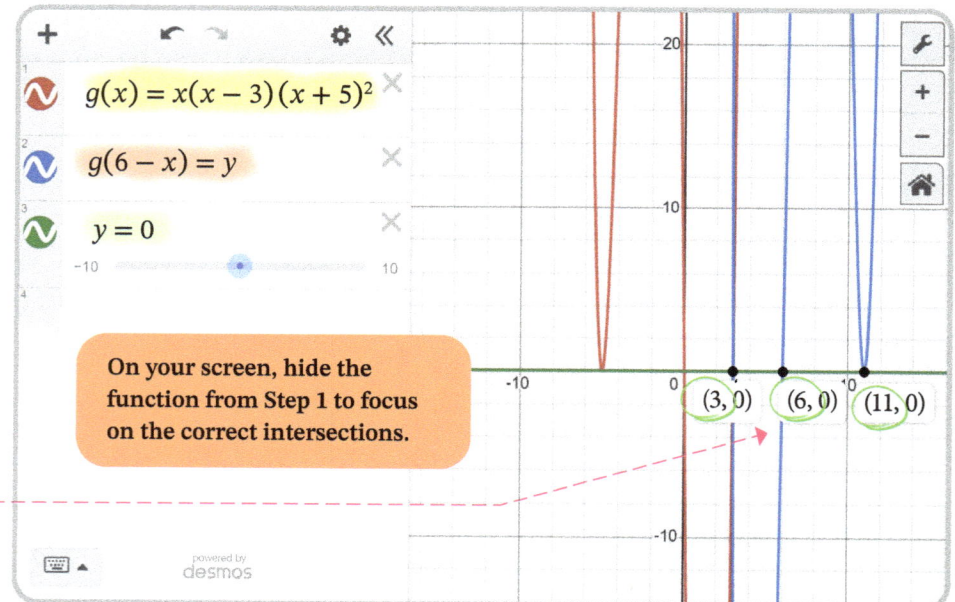

On your screen, hide the function from Step 1 to focus on the correct intersections.

At first glance, this question looks like it might work with Technique #1. But Technique #1 won't reveal all the solutions, while Technique #6 will—so use Technique #6 here.

Exercise #49

$8\sqrt[5]{3^5 x^{20}} \cdot \sqrt[6]{2^6 x^{24}} = 48x^b$

In the equation above, b is a positive constant and $x > 1$. Which of the following is the value of b?

(A) 7   (B) 8   (C) 9   (D) 10

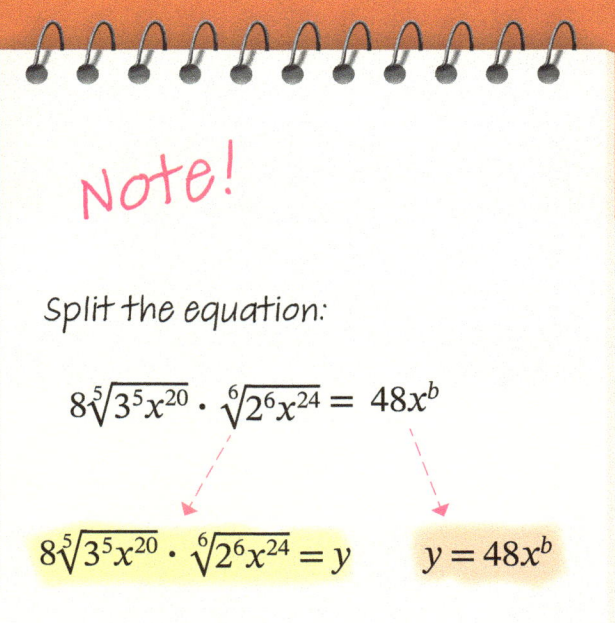

Note!

Split the equation:

$8\sqrt[5]{3^5 x^{20}} \cdot \sqrt[6]{2^6 x^{24}} = 48x^b$

$8\sqrt[5]{3^5 x^{20}} \cdot \sqrt[6]{2^6 x^{24}} = y$ $y = 48x^b$

How to Solve

Step 1 Enter the two split equations.

Step 2 Add a slider for b and adjust it until the graphs overlap.

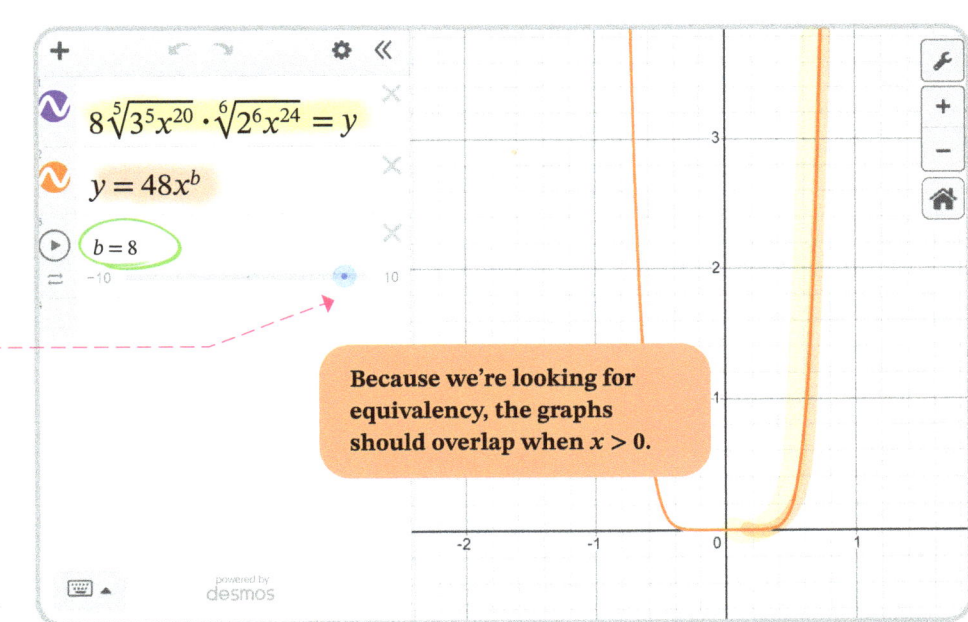

Because we're looking for equivalency, the graphs should overlap when $x > 0$.

The answer is B.

That little 5 tucked into the root sign means the fifth root, not an exponent. To enter it, go to [functions] on the Desmos Keypad and select [∜] or type **nthroot** and Desmos will convert it automaticlly.

$$2x^2 + 30x + c = 0$$

In the equation above, c is a constant. What is the smallest integer value of c for which the equation has no real solutions?

_____

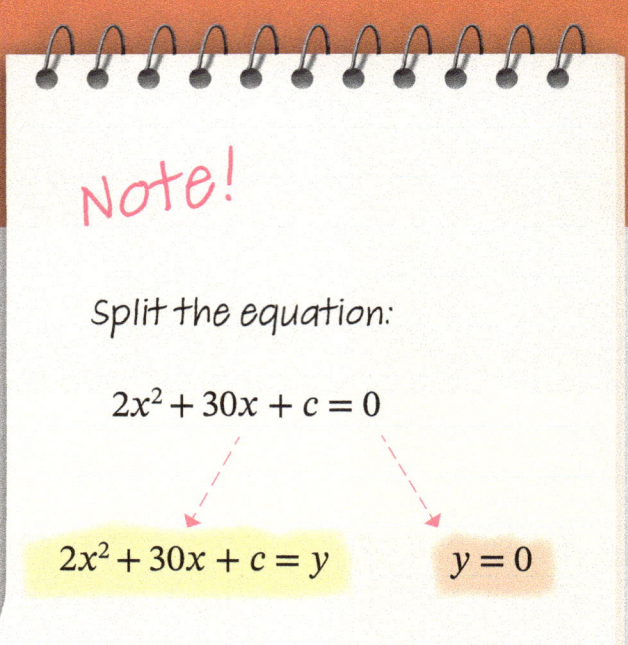

Note!

Split the equation:

$$2x^2 + 30x + c = 0$$

$$2x^2 + 30x + c = y \qquad y = 0$$

How to Solve

Step 1 Enter the two split equations. Add a slider for c.

Step 2 Adjust the slider and expand its range to find the smallest integer value where the parabola and line no longer intersect.

The answer is 113.

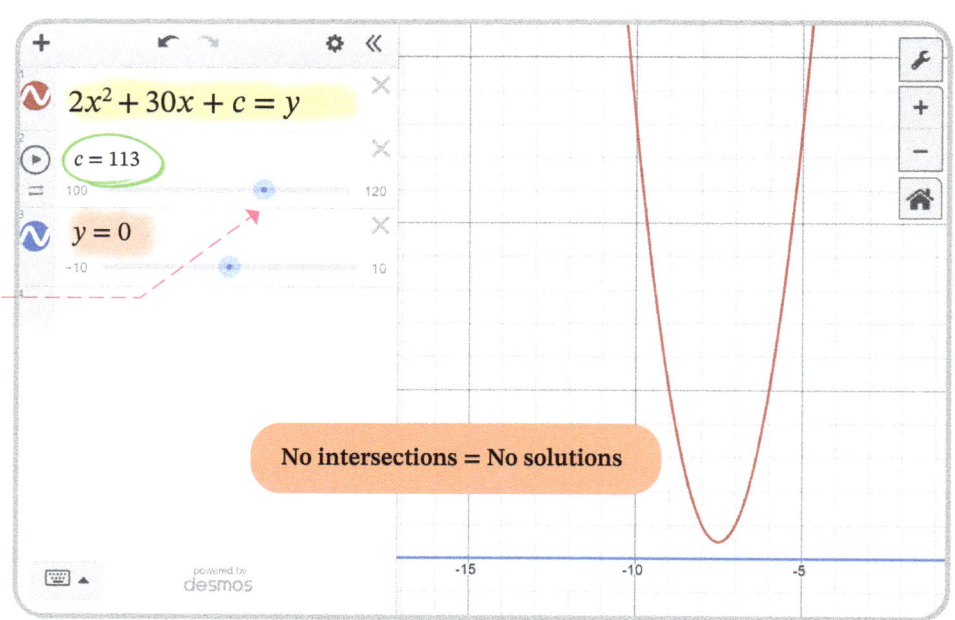

$2x^2 + 30x + c = y$

$c = 113$

100   120

$y = 0$

−10   10

No intersections = No solutions

powered by
desmos

−15   −10   −5

And just like that, you've learned all the techniques.

Technique #6
PRACTICE PROBLEMS

Problem 1

$-2x^2 + 4x - 56 = 0$

How many real solutions are there to the given equation?

A) Zero
B) One
C) Two
D) Infinitely many

Problem 2

$-x^2 - 3x - d = 0$

In the equation above, d is a constant. What is the smallest integer value of d for which the equation has no real solutions?

_____

Hint: Desmos adds a slider when you enter $y = 0$, but don't adjust it. Instead, adjust the slider for d.

Problem 3

$x^2 - 4x - c = 3$

In the given equation, c is a constant. If the equation has exactly one solution, what is the value of c?

A) -3
B) 3
C) 5
D) -7

Hint: Do not adjust the slider for $y = 3$.

Problem 4

The function f is defined by $f(x) = (x - 9)(x - 15)(x - 2)^2$. If $f(x + 7) = 0$, what is the sum of all possible values of x?

A) -3
B) 2
C) 5
D) 8

Problem 5

$$3x^2 - 48x + c = 0$$

In the equation above, c is a constant. What is the value of c for which the equation has exactly one real solution?

_____

Hint: Zoom out!

Problem 6

$$x^2 - 4x + 8 = 0$$

How many real solutions are there to the given equation?

A) Zero
B) One
C) Two
D) Infinitely many

Problem 7

$$2\sqrt[3]{4^{\frac{3}{2}}x^3} \cdot \sqrt[4]{25^2 x^4} = bx^2$$

In the equation above, b is a positive constant and $x > 1$. Which of the following is the value of b?

_____

Problem 8

$$4x + 5ax = 10$$

In the equation above, a is a constant. If the equation has no solutions, what is the value of a?

A) $-\dfrac{5}{4}$

B) $\dfrac{5}{4}$

C) $-\dfrac{4}{5}$

D) $\dfrac{4}{5}$

Problem 9

The function f is defined by
$f(x) = (x + 8)(x - 20)(x + 3)^2$. If
$f(13 - x) = 0$, what is the sum of all
possible values of x?

_____

Problem 10

$$\frac{a}{2}x - b = \frac{3}{5}x - 5$$

In the equation above, a and b are
constants. If the equation has no solutions
and $b \neq 5$, what is the value of a?

_____

Technique #6
PRACTICE PROBLEMS

Answers		References
1	A	*Exercise #47*
2	3	*Exercise #50*
3	D	*Exercise #50*
4	C	*Exercise #48*
5	192	*Exercise #50*
6	A	*Exercise #47*
7	20	*Exercise #49*
8	C	*Exercise #46*
9	30	*Exercise #48*
10	1.2 or $\frac{6}{5}$	*Exercise #46*

Notes

Practice
Makes
Perfect

Practice Makes Perfect

Welcome to the final section. Here, you'll find 10 problem sets, each with 10 questions designed to strengthen the strategies you've learned in *SAT Math Made Visual.*

To-Do List

* Practice

* Spot patterns

* See math visually

* Take a walk with your pet

Practice Makes Perfect
PROBLEM SET #1

Problem 1

What is the equation of a line that passes through $(15, -1)$ and is perpendicular to the line $-10x + 2y = 72$?

A) $-5x - y = 10$

B) $5x + y = -10$

C) $x + 5y = -10$

D) $x + 5y = 10$

Problem 2

$2jx - 5y = 0$
$5x + y = 10$

In the system of equations above, j is a constant, and x and y are variables. If the system has no solutions, what must be the value of j?

A) -12.5
B) 25
C) 12.5
D) -25

Problem 3

$$\sqrt{3x + 4} = \sqrt{(x + 2)^2}$$

In the equation above, what is the smallest value of x?

_____

Problem 4

A school fundraiser sells smoothies and sandwiches. Buying 3 smoothies and 4 sandwiches costs \$28.50, while buying 5 smoothies and 2 sandwiches costs \$24.40. What is the price of one smoothie?

A) \$1.82

B) \$2.43

C) \$2.90

D) \$4.95

Practice Makes Perfect
PROBLEM SET #1

The result of increasing the quantity k by 1,600% is 544. What is the value of k?

_____

Hint: One way to write the equation: $x + 1600\%$ of $x = 544$.

Problem 6

A quadratic function passes through $(4, 41)$ and $(-4, 97)$ and is written in the form $y = ax^2 + bx + 5$. What is the value of b?

_____

Problem 7

Given the function $g(x) = 3x^2 + 47x - 27$, which table shows the correct x values and their corresponding $g(x)$ values?

A)

x	$g(x)$
-2	-71
1	23
6	283

B)

x	$g(x)$
-1	-71
2	79
5	363

C)

x	$g(x)$
2	23
6	363
10	639

D)

x	$g(x)$
-1	-71
5	283
10	743

Problem 8

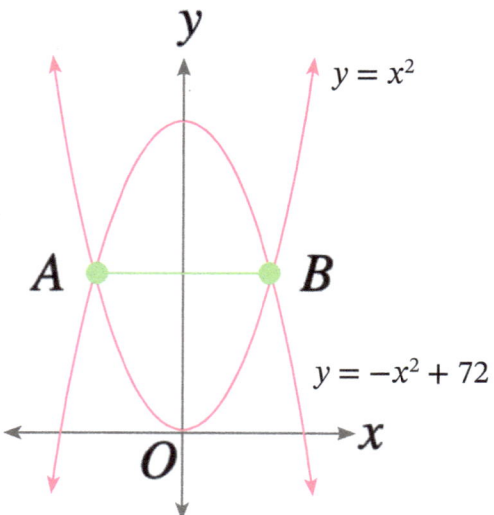

Note: Figure not drawn to scale.

The figure above shows the graphs of $y = x^2$ and $y = -x^2 + 72$. What is the length of segment AB?

A) 24
B) 16
C) 12
D) 8

Problem 9

If the graph of $6x + 3y = 12$ is translated up five units, what is the x-intercept of the translated line?

A) $(5, 0)$
B) $(4.5, 0)$
C) $(4, 0)$
D) $(3.5, 0)$

Problem 10

$$\frac{4x^2}{2x + 3b} = \frac{8x^2}{4x + 42}$$

In the equation above, b is a constant and $x > 0$. What is the value of b?

Practice Makes Perfect
PROBLEM SET #1

Answers		References
1	D	*Exercise #30*
2	A	*Exercise #28*
3	−1	*Exercise #2*
4	C	*Exercise #22*
5	32	*Exercise #4*
6	−7	*Exercise #43*
7	D	*Exercise #10*
8	C	*Exercise #19*
9	B	*Exercise #41*
10	7	*Exercise #49*

Notes

Practice Makes Perfect
PROBLEM SET #2

Problem 1

A line passes through $(0, 0)$, $(2, n)$, and $(n, 18)$. What is one possible value of n?

A) 3

B) 6

C) 9

D) 12

Problem 2

$$\frac{5x(x - 21) - 2(x - 21)}{2x - 42}$$

Which expression is equivalent to the expression above, where $x > 21$?

A) $\frac{5x - 2}{2}$

B) $5x - 1$

C) $5x^2 - 105x - 2$

D) $\frac{5x^2 - 105x - 21}{2x - 42}$

Problem 3

$|18 - x| = 4$
$|x - 12| = 10$

For what value of x are the equations above true?

Hint: Find the vertical line that works for both equations. Hide the graph that's on top to reveal the overlapping line underneath.

Problem 4

The equation $v(t) = 10{,}000(0.90)^t$ gives the estimated value of a car (in dollars), where t is the number of years since the car was purchased. The value of the car decreases by $p\%$ each year. What is the value of p?

Problem 5

The graph of $2x^2 + 2x + 2y^2 + 2y = 199$ is a circle. What is the diameter of the circle?

Hint: Do not mistake the y-intercepts as the minimum and maximum points.

Problem 6

$$y = a(x + 1)(x - 3)$$

In the quadratic equation above, a is a nonzero constant. The graph of the equation is a parabola with vertex (h, k). Which of the following is equivalent to k?

A) $-a$
B) $-3a$
C) $-4a$
D) $-6a$

Hint: Set a to any nonzero value with the slider. Then plug that value of a into each answer choice and see which one matches the y value of the vertex.

Problem 7

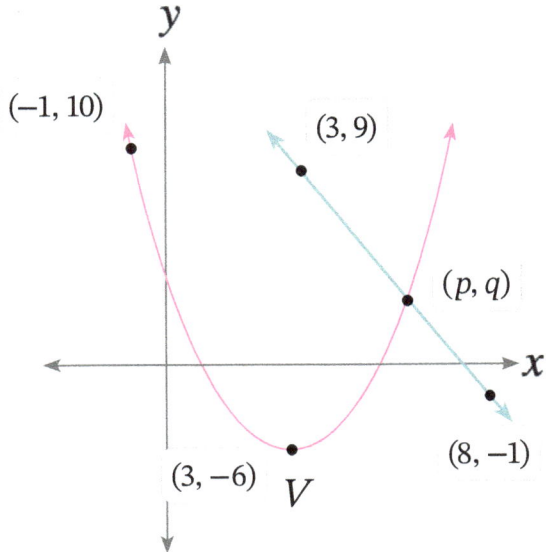

Note: Figure not drawn to scale.

The graph above shows one of the two points where a line and parabola intersect, labeled (p, q). V is the vertex of the parabola. What is the value of p if $p > 0$?

Hint: For the parabola, create a quadratic regression model in vertex form. Do not enter (p, q) in the table for either the line or the parabola.

Practice Makes Perfect
PROBLEM SET #2

Problem 8

$h(t) = -16t^2 + 64t$

The equation above represents the height $h(t)$ of a baseball t seconds after it is hit straight into the air at a speed of 64 feet per second. After how many seconds will the baseball hit the ground?

A) 4.0

B) 3.2

C) 2.8

D) 2.0

Problem 9

$-4x^2 + 50x + c = 0$

In the equation above, c is a constant. What is the largest integer value of c for which the equation has no real solutions?

_____

Hint: Use Technique #6. Zoom out to see the graphs better.

Problem 10

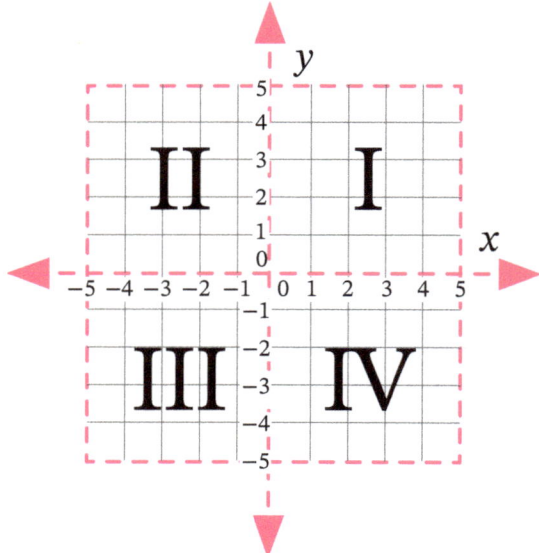

$y < x - 2$

$y \le \frac{x}{2} - 1$

Which of the following quadrants contains no solutions to the system of inequalities?

A) Quadrant I

B) Quadrant II

C) Quadrant III

D) Quadrant IV

Practice Makes Perfect
PROBLEM SET #2

Answers		References
1	B	*Exercise #14*
2	A	*Exercise #29*
3	22	*Exercise #3*
4	10	*Exercise #11*
5	20	*Exercise #12*
6	C	*Exercise #16*
7	6	*Exercise #45*
8	A	*Exercise #18*
9	−157	*Exercise #50*
10	B	*Exercise #24*

Notes

Practice Makes Perfect
PROBLEM SET #3

Problem 1

The function f is defined by $f(x) = 2x - 4$, and for some number b, $f(4) - f(b) = f(1)$. What is the value of b?

A) -2
B) 2
C) 3
D) 5

Hint: Replace b with x.

Problem 2

$$3x + 7y = 20$$
$$12x = 80 - ay$$

What is the value of a for which the system of equations has infinitely many solutions?

_____

Problem 3

$$36^{\frac{5}{2}} = \sqrt[4]{6^x}$$

What is the value of x in the equation above?

A) 20
B) 10
C) 5
D) 2

Problem 4

A plumber charges an initial fee of d dollars and an additional m dollars per hour worked. A customer pays $220 for a 2-hour job and $340 for a 4-hour job. How much will a 10-hour job cost?

_____

Practice Makes Perfect
PROBLEM SET #3

Problem 5

$$\frac{3}{49}y = 2x^2 - 40x - 3$$
$$y = 2x^2 - 80x + 7$$

In the system of equations above, one solution is (a, b). Which of the following is the closest value of a?

A) 2.563
B) 1.544
C) 0.076
D) 3.145

Problem 6

Lines p and q are perpendicular. Line p has a slope of 5. Line q passes through $(6, 7)$ and $(16, c)$. What is the value of c?

A) 4
B) 5
C) 6
D) 7

Problem 7

$$2x + b = 4x^2 - 2x + 5$$

In the equation above, b is a constant. For what value of b is there only one real solution to the equation?

A) 4
B) 2
C) $\frac{1}{2}$
D) $\frac{1}{4}$

Practice Makes Perfect
PROBLEM SET #3

A line passes through the points $(-3a, 40)$, $(-a, 34)$, and $(a, 28)$. Which equation represents the line?

A) $3x + ya = 31$

B) $3x + ya = 31a$

C) $ax + 3a = 31a$

D) $ax + 3a = 31$

Hint: Enter the points in the Expression List and accept a slider for a. Then enter each answer choice to see which graph passes through the points. At first, the graph for answer choice A might seem to work—but when you move the slider, it no longer passes through all the points, so it isn't the answer.

Problem 9

$2x - x^2 + 4y - y^2 = -4$

The equation above represents a circle. What is the radius of the circle?

A) 9

B) 6

C) 4

D) 3

Problem 10

The function h is defined by $h(x) = x(x - 1)(x + 3)^2$. If the value of $h(5 - d) = 0$ and d is a constant, what is the sum of all possible values of d?

A) 12

B) 13

C) 17

D) 25

Hint: Use Technique #6. Rewrite and split the second function:
$h(5 - x) = y$
$y = 0$

Practice Makes Perfect
PROBLEM SET #3

Answers		References
1	D	*Exercise #21*
2	28	*Exercise #27*
3	A	*Exercise #6*
4	700	*Exercise #38*
5	C	*Exercise #26*
6	B	*Exercise #40*
7	A	*Exercise #50*
8	B	*Exercise #14*
9	D	*Exercise #12*
10	C	*Exercise #48*

Notes

_____

_____

_____

_____

_____

_____

_____

_____

_____

_____

_____

_____

_____

_____

Practice Makes Perfect
PROBLEM SET #4

Problem 1

$3x^2 - 12x + c = 0$

In the equation above, c is a constant. What is the largest integer value of c for which the equation has two real solutions?

_____

Hint: After splitting the equation, make sure to adjust the slider for c, not y.

Problem 2

The function h is defined by $h(x) = 10x - 5$. If $2h(m + 2) = 100$, what is the value of m?

A) 5
B) 4.5
C) 4
D) 3.5

Problem 3

x	y
3	−3
5	7

The data points in the table represent solutions to the line $y = mx + b$, where m and b are constants. What is the value of $m + b$?

_____

Problem 4

The function h is defined by $h(x) = (x - 12)(k + x)$, where k is a constant. The graph of $y = h(x)$ passes through $(6, 0)$. What is the value of $h(0)$?

_____

Hint: The answer to $h(0)$ is the y value of the y-intercept.

Practice Makes Perfect
PROBLEM SET #4

Problem 5

$4a - 4b = 10$

$7a + 3b = 17$

How many solutions are there to the system of equations above?

A) Zero
B) One
C) Two
D) Infinitely many

Problem 6

Line n passes through $\left(\frac{1}{4}, \frac{3}{8}\right)$ and $(a, 1)$ and has a slope of $\frac{1}{6}$. What is the value of a?

Problem 7

$25x^2 - 40x = 2y + 100$

$30x^2 + 43x = y$

The solution to the given system of equations is (x, y). What is the smallest value of x, rounded to the nearest tenth?

A) 1.2
B) −1.2
C) 2.4
D) −2.4

Hint: The question asks for the smallest value, not just a correct one.

Problem 8

$x^2 + 5x - 5 = 0$

What is the sum of all the values of x that satisfy the equation above?

A) 5
B) −5
C) $5 + \sqrt{5}$
D) $5 - \sqrt{5}$

Problem 9

$x^2 + y^2 - 6x + 4y = c$

The equation above represents a circle with diameter endpoints (3, 4) and (3, −8). What is the value of c?

A) 23
B) 18
C) 9
D) 4

Problem 10

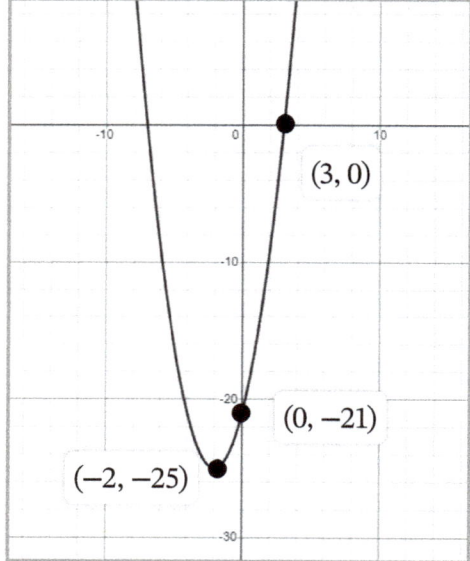

(3, 0)

(0, −21)

(−2, −25)

The graph of the quadratic function is shown above. What is the value of y when $x = 8$?

Practice Makes Perfect
PROBLEM SET #4

Answers		Reference
1	11	*Exercise #50*
2	D	*Exercise #21*
3	−13	*Exercise #36*
4	72	*Exercise #15*
5	B	*Exercise #27*
6	4	*Exercise #40*
7	D	*Exercise #26*
8	B	*Exercise #1*
9	A	*Exercise #13*
10	75	*Exercise #42*

Notes

_____

_____

_____

_____

_____

_____

_____

_____

_____

_____

_____

_____

_____

Practice Makes Perfect
PROBLEM SET #5

Problem 1

$7 < |3x - 14| < 16$

What is the greatest possible integer value of x that satisfies the inequality?

Problem 2

The function g is defined as $g(x) = a\sqrt{2x + b}$, where a and b are constants. The graph of $y = g(x)$ passes through point $(5, 0)$. If all solutions have a positive y value, what must be true?

A) $a > b$

B) $a < b$

C) $b > 10$

D) $b < -10$

Problem 3

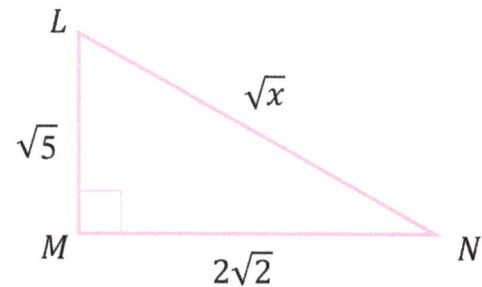

Note: Figure not drawn to scale.

In the right triangle LMN above, leg LM is $\sqrt{5}$ cm and MN is $2\sqrt{2}$ cm. If the hypotenuse can be written as $\sqrt{x}$ cm, what is x?

Problem 4

A 45−45−90 triangle has a perimeter of $64 + 64\sqrt{2}$ units. What is the length of one leg of the triangle?

A) 64

B) $64\sqrt{2}$

C) $32\sqrt{2}$

D) 32

Practice Makes Perfect
PROBLEM SET #5

Problem 5

$$\frac{3x^2}{2x + 2a} = \frac{6x^2}{4x + 30}$$

In the equation above, a is a constant and $x > 0$. What is the value of a?

A) -7.5
B) 7.5
C) 15
D) -15

Problem 6

$$4x^2 - 35x - 100 = 0$$

How many real solutions does the given equation have?

A) Zero
B) One
C) Two
D) Infinitely many

Problem 7

The graph of a line passes through $(0, a + 10)$, $(3, a + 19)$, and $(10, 2a - 2)$. What is one possible value of a?

A) 20
B) 35
C) 39
D) 42

Problem 8

$$f(x) = 1.46^{\frac{x}{3}}$$

The function f is depicted above. Its equation can be expressed as

$$f(x) = \left(1 + \frac{p}{100}\right)^x$$

where p is a constant. Which of the following values is closest to p?

A) 13.5
B) 14
C) 14.5
D) 15

Hint: Zoom out!

Problem 9

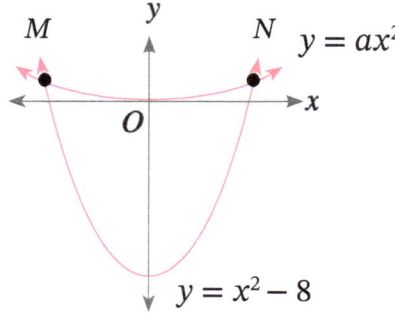

M y N $y = ax^2$

$y = x^2 - 8$

Note: Figure not drawn to scale.

The figure above shows the graphs of $y = ax^2$ and $y = x^2 - 8$. The distance between points M and N is 8. W is the value of a?

A) $\frac{2}{3}$

B) $\frac{1}{2}$

C) $\frac{1}{3}$

D) $\frac{1}{4}$

Hint: Create a slider, but instead of adjusting it, type in each answer choice for a and see which one makes the distance between M and N equal to 8.

Problem 10

x	$h(x)$
-3	7
-2	a
-1	4
1	1
2	c

The table above gives values of the function h for several values of x. If the graph of $h(x)$ is a line, which of the following equals $a + c$?

Hint: You only need to enter two coordinate points in the table to find a linear regression. Do not use points $(-2, a)$ or $(2, c)$.

Practice Makes Perfect
PROBLEM SET #5

Answers

1 9

2 A

3 13

4 C

5 B

6 C

7 D

8 A

9 B

10 5

References

Exercise #25

Exercise #15

Exercise #7

Exercise #8

Exercise #49

Exercise #47

Exercise #14

Exercise #35

Exercise #19

Exercise #38

Notes

_____

_____

_____

_____

_____

_____

_____

_____

_____

_____

_____

_____

_____

_____

Practice Makes Perfect
PROBLEM SET #6

Problem 1

$10x - 15y = 20$

$\frac{1}{2}x + ky = 15$

In the system of equations above, k is a constant. For what value of k are there no solutions to the system?

A) $-\frac{4}{3}$

B) $-\frac{3}{4}$

C) $\frac{4}{3}$

D) $\frac{3}{4}$

Problem 2

$f(x) = x^2 + \frac{17}{9}x - \frac{2}{9}$

If $f(x) = 0$ and x is positive, what is the value of $45x$?

_____

Hint: Use $\frac{1}{9}$ for 0.11111 or round the answer at the end.

Problem 3

The function f is defined as $f(x) = 7x^3$. The graph of $h(x)$ is the result of shifting the graph of $y = f(x)$ to the right 3 units. Which equation defines the function h?

A) $h(x) = 7x^3 - 3$

B) $h(x) = 7x^3 + 3$

C) $h(x) = 7(x^3 - 9x^2 + 27x - 27)$

D) $h(x) = 7(x^3 + 9x^2 + 27x + 27)$

Problem 4

A circle is centered at $(-3, 0)$ and contains the point $\left(-1, \frac{8}{3}\right)$ on its circumference. Which of the following is an equation of the circle?

A) $(x - 3)^2 + y^2 = \frac{10}{3}$

B) $(x - 3)^2 + y^2 = \frac{100}{9}$

C) $(x + 3)^2 + y^2 = \frac{10}{3}$

D) $(x + 3)^2 + y^2 = \frac{100}{9}$

Practice Makes Perfect
PROBLEM SET #6

Problem 5

$(a^4b^5c^{-2})(ab^3c^{-3})$

Which expression is equivalent to the expression above, where a, b, and c are positive?

A) $a^4b^{15}c^6$
B) $a^5b^{15}c$
C) $a^5b^8c^6$
D) $a^5b^8c^{-5}$

Hint: When $c = 1$, answer choices C and D both overlap the given expression. As you move the slider, only the correct choice continues to overlap.

Problem 6

$2^{3d+1} = 16 \cdot 2^{d-2}$

What is the value of d in the equation above?

Problem 7

A hot air balloon is launched from an initial height of 15 meters above the ground. Until the balloon descends to the ground, its height $h(t)$, in meters, t seconds after it was launched, is given by the following function, where c is a constant:

$h(t) = 40 - (2t - c)^2$

If the balloon reached its maximum height of 40 meters exactly 2.5 seconds after it was launched, what was its height 4 seconds after it was launched?

A) 35
B) 31
C) 15
D) 0

Practice Makes Perfect
PROBLEM SET #6

Let the function g be defined by $g(x) = x^2 + 12$. For what positive number n is it true that $g(3n) = 3g(n)$?

_____

In the equation $4x - 8px = 16$, for what value of p does the equation have no solutions?

A) $\frac{1}{2}$

B) 1

C) $\frac{1}{4}$

D) $\frac{2}{3}$

Hint: Typing a fraction for p temporarily removes the slider. Typing a whole number or decimal will bring it back.

For $x > 0$, the function h can be expressed as $h(x) = 350\%$ of x. Which of the following could describe this function?

A) Decreasing linear
B) Decreasing exponential
C) Increasing linear
D) Increasing exponential

Hint: Increasing functions rise as x values increase. Decreasing functions fall as x values increase. Linear functions form straight lines. Exponential functions curve upward or downward, getting steeper as they go.

Practice Makes Perfect
PROBLEM SET #6

Answers

1	B
2	5
3	C
4	D
5	D
6	.5 or $\frac{1}{2}$
7	B
8	2
9	A
10	C

References

Exercise #28

Exercise #20

Exercise #41

Exercise #12

Exercise #29

Exercise #6

Exercise #18

Exercise #21

Exercise #46

Exercise #4

Notes

_____

_____

_____

_____

_____

_____

_____

_____

_____

_____

_____

_____

_____

Problem 1

Given the function $g(x) = 3x^2 + 12x - 117$, which table shows the correct x values and their corresponding $g(x)$ values?

A)

x	$g(x)$
−2	−129
1	−102
7	114

B)

x	$g(x)$
−2	−129
2	−81
3	−21

C)

x	$g(x)$
−3	−126
0	−117
10	234

D)

x	$g(x)$
−1	−126
0	−117
1	−81

Problem 2

136, 555, 1001, 1002, 5, 33, 56, 43, 424, 1

What is the median of the data set shown?

_____

Hint: Expand the Expression List by dragging its boundary to see all the entries.

Problem 3

x	$f(x)$
2	3
1	12

The table above shows points for the quadratic function f. If the function can be written as $f(x) = -2x^2 + bx + c$, what is the sum of b and c?

A) −20
B) −3
C) 14
D) 17

Problem 4

Line j passes through $(5, 1)$ and is perpendicular to the line $5x + 2y = 10$. What is the equation of line j?

A) $2x - 5y = -5$

B) $2x + 5y = 5$

C) $2x + 5y = -5$

D) $2x - 5y = 5$

Problem 6

$x^2 - 3x + 5 = 0$

How many real solutions are there to the given equation?

A) Zero
B) One
C) Two
D) Infinitely many

Problem 5

Which of the following functions below has/have a minimum value at -4?

I. $f(x) = -5(4)^x - 4$
II. $h(x) = -4(5)^x$

A) I only
B) II only
C) I and II
D) Neither I nor II

Problem 7

A parabola has a vertex at $(-3, -8)$ and passes through the point $(-1, 4)$. What is the y-intercept of the parabola?

A) $(0, 18)$
B) $(0, 19)$
C) $(0, 20)$
D) $(0, 21)$

Practice Makes Perfect
PROBLEM SET #7

Problem 8

When the equation $y = |x + 15|$ is graphed, it intersects the line $y = 15$ at two points, P and Q. What is the length of segment PQ?

A) 60

B) 30

C) 15

D) 10

Problem 9

Which expression is equivalent to $x^{\frac{1}{10}}$ when $x > 0$?

A) $\sqrt[2]{x^{20}}$

B) $\sqrt{x^{10}}$

C) $\sqrt[4]{x^{\frac{2}{5}}}$

D) $\sqrt[5]{x^{\frac{2}{3}}}$

Problem 10

The graph of $4x - 3y = 15$ is translated down 3 units. What is the x-intercept of the new line?

A) $(6, 0)$

B) $(5.5, 0)$

C) $(5, 0)$

D) $(4.5, 0)$

Practice Makes Perfect
PROBLEM SET #7

Answers

1	A
2	96
3	C
4	D
5	D
6	A
7	B
8	B
9	C
10	A

References

Exercise #10

Exercise #5

Exercise #43

Exercise #31

Exercise #16

Exercise #47

Exercise #44

Exercise #19

Exercise #34

Exercise #41

Notes

_____

_____

_____

_____

_____

_____

_____

_____

_____

_____

_____

_____

_____

Practice Makes Perfect
PROBLEM SET #8

Problem 1

A 30−60−90 degree triangle has a perimeter of 90 units. What is the length of its shortest side? Round to the nearest whole number.

_____

Problem 2

The number of mold spores in a sealed container triples every day. There are 5,000 mold spores at the start of an observation. Which of the following equations represents the number of mold spores, $f(x)$, in the container x days after the observation begins?

A) $f(x) = 5{,}000 + 3^x$

B) $f(x) = \frac{1}{3}(5{,}000)^x$

C) $f(x) = 5{,}000(3)^x$

D) $f(x) = 3(5{,}000)^x$

Hint: Desmos can't generate more than one Function Table for $f(x)$. Replace $f(x)$ with y for each answer choice.

Problem 3

$$y = 2x + 4$$

$$y = x^2 - 8x - a$$

In the system of equations above, a is a constant. The system has exactly one real solution. What is the value of a?

A) −29
B) −5
C) 5
D) 14

Problem 4

Line j passes through (3, 2) and is perpendicular to a line with a slope of $\frac{1}{5}$. What is the y-intercept of line j?

A) (0, −17)
B) (0, −13)
C) (0, 13)
D) (0, 17)

Problem 5

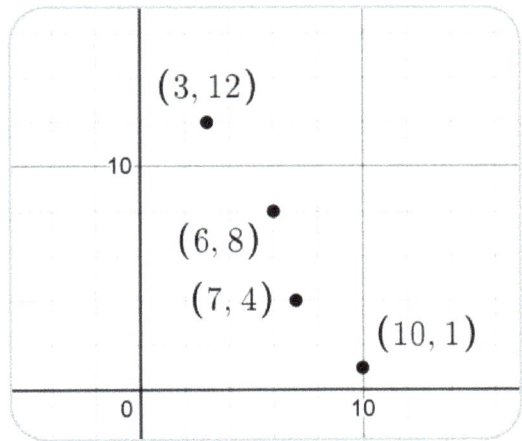

Which of the following equations best represents the trend shown by the data above?

A) $y - 2x = 19$

B) $y + 2x = 19$

C) $y + 2x = -19$

D) $y - 2x = -19$

Problem 6

24 is $p\%$ of 120% of 50. What is the value of p?

Problem 7

The median of $x - 1$, $x - 2$, and $\frac{3x + 7}{4}$ is 6. What is the value of x?

A) 10
B) 9
C) 8
D) 7

Problem 8

A linear equation passes through $(0, -24)$ and $(16, 40)$. A quadratic function passes through $(-2, 15.5)$, $(0, 3)$, and $(10, 23)$. Both functions intersect at (x, y). What could be the value of y?

A) 12
B) 8
C) 4
D) 2

Problem 9

Two identical rectangular prisms each have a height of 50 cm. Each prism has a square base and a surface area of M cm². When their square bases are glued together, the resulting prism has a surface area of $\frac{23}{15} M$ cm². What is the side length, in cm, of each square base?

_____

Hint: Write the equations as follows, with x representing the side length and y replacing M:

$y = 2x^2 + 4(50x)$

$\frac{23}{15} y = 2x^2 + 4(100x)$

Combine the equations to get the following:

$\frac{23}{15}(2x^2 + 4(50x)) = 2x^2 + 4(100x)$

Problem 10

$3x + 4y = -40$

For each real number r, which of the following coordinate points always lies on the graph of the equation above?

A) $\left(\frac{r}{4}, 3r\right)$

B) $\left(\frac{r}{2}, -\frac{3r}{8} - 10\right)$

C) $\left(\frac{2r}{5}, r\right)$

D) $\left(\frac{r}{8}, 3r + 10\right)$

Hint: Enter the equation and each answer choice. Accept the slider for r, then look for the point that always stays on the line as you move the slider.

Practice Makes Perfect
PROBLEM SET #8

Answers		References
1	19	*Exercise #8*
2	C	*Exercise #11*
3	A	*Exercise #23*
4	D	*Exercise #40*
5	B	*Exercise #9*
6	40	*Exercise #4*
7	D	*Exercise #5*
8	B	*Exercise #45*
9	87.5 or $\frac{175}{2}$	*Exercise #26*
10	B	*Exercise #14*

Notes

_____

_____

_____

_____

_____

_____

_____

_____

_____

_____

_____

_____

_____

_____

Problem 1

$$-2x + 10a = (2x + 10)(3 - 4x)$$

What is the smallest integer value of a for which there are no solutions to the equation above?

_____

Problem 2

A quadratic function has a vertex of $(1, 9)$ and an x-intercept of $(4, 0)$. What is the y-intercept of the function?

A) $(0, 7)$
B) $(0, 7.5)$
C) $(0, 8)$
D) $(0, 8.5)$

Problem 3

$$f(x) = x^2 - 4x + 9$$

The given equation defines the function f. If $g(x) = 2f(x - 3)$, what is the vertex of $g(x)$?

A) $(2, 5)$
B) $(5, 10)$
C) $(2, 3)$
D) $(5, 5)$

Problem 4

The function h is defined by $h(x) = -9x^2 + 12x + 16$. If $h(x) = 20$, what is the value of $60x$?

_____

Practice Makes Perfect
PROBLEM SET #9

Problem 5

64 is $x\%$ of x. What is x?

A) 74
B) 76
C) 78
D) 80

Problem 6

A parabola goes through $(0, -9)$, $(3, 12)$ and $(-5, -4)$. If its equation is written as $y = ax^2 + bx + c$, what is the value of $a + b + c$?

A) -4
B) 4
C) 9
D) -9

Problem 7

The function $g(x) = 100{,}000(2)^{\frac{x}{320}}$ gives the number of bacteria in a population x minutes after an initial observation. How many minutes does it take the number of bacteria in the population to double?

$$\rule{4cm}{0.4pt}$$

Problem 8

$$2x - \frac{y - 4}{7} = 14$$

$$\frac{x}{2} + \frac{y + 3}{3} = 26.5$$

In the system of equations above, what is the value of x?

A) 12
B) 11.5
C) 11
D) 10.5

Problem 9

x	$f(x)$
-1	25
1	9
8	16
10	36

The table above shows values of the function f for several values of x, where f can be defined by $f(x) = (x - b)^2$. What is the value of b?

_____

Problem 10

$$6\sqrt[4]{2^4 x^{24}} \cdot \sqrt[9]{3^9 x^{18}} = 36x^b$$

In the equation above, b is a positive constant and $x > 1$. Which of the following is the value of b?

A) 8

B) 8.5

C) 9

D) 9.5

Practice Makes Perfect
PROBLEM SET #9

	Answers		References
1	7		*Exercise #50*
2	C		*Exercise #44*
3	B		*Exercise #17*
4	40		*Exercise #20*
5	D		*Exercise #4*
6	A		*Exercise #42*
7	320		*Exercise #11*
8	C		*Exercise #19*
9	4		*Exercise #15*
10	A		*Exercise #49*

Notes

Problem 1

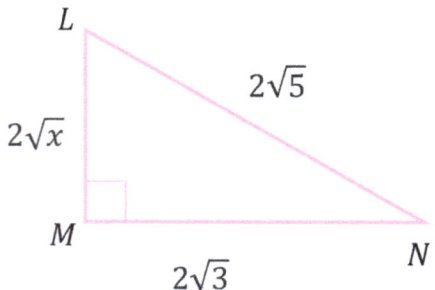

L

$2\sqrt{5}$

$2\sqrt{x}$

M

N

$2\sqrt{3}$

Note: Figure not drawn to scale.

In the right triangle LMN above, the length of leg LM is $2\sqrt{x}$ cm and the length of MN is $2\sqrt{3}$ cm. If the hypotenuse LN can be written as $2\sqrt{5}$ cm, what is x?

Problem 2

The average of $\frac{x+1}{6}$, $x - 34$, and $\frac{2x}{35}$ is 3. What is x?

Problem 3

$$300x^2 - 41x = y + 620$$
$$250x^2 + 4x = y$$

The solution to the given system of equations is (a, b). What is the positive value of a?

A) 16
B) 12
C) 10
D) 4

Problem 4

$$(x^{-2}y^3z^{-4})(xy^{-1}z^5)$$

Which expression is equivalent to the expression above, where $x, y,$ and z are positive?

A) $x^{-1}y^2z$
B) $x^{-2}y^2z^{-1}$
C) $x^{-3}y^2z$
D) $x^{-2}y^3z^{-1}$

Problem 5

The function $y = 1,200(2)^{\frac{x}{30}}$ gives the number of cells in a culture, where x is the number of minutes since the initial measurement. At $x = 0$, there are 1,200 cells. How many minutes does it take for the number of cells to double?

_____

Problem 6

$$h(t) = -16t^2 + 64t + 16$$

The equation above represents the height, $h(t)$, in feet, of an object t seconds after it is hit straight into the air at a speed of 64 feet per second from a platform 16 feet high. What is the maximum height the object will reach?

A) 2
B) 4
C) 80
D) 100

Problem 7

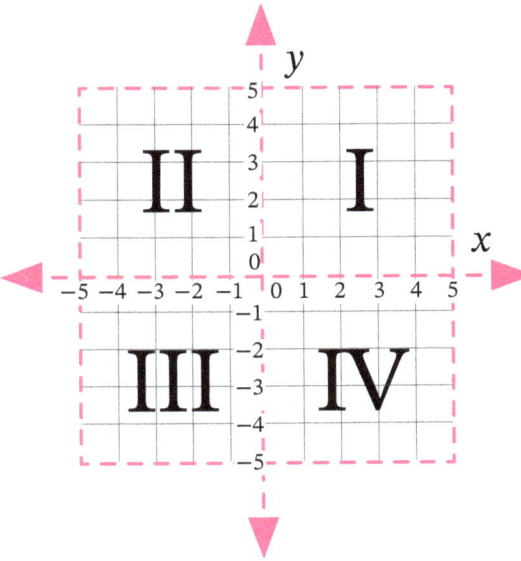

$$y > \frac{2}{3}x$$
$$y \geq 2x + 5$$

If the system of inequalities above is graphed, which of the following quadrants contains no solutions to the system?

A) Quadrant I
B) Quadrant II
C) Quadrant III
D) Quadrant IV

Practice Makes Perfect
PROBLEM SET #10

Problem 8

$x + 4y = 10$
$2x + by = 15$

In the given pair of equations, b is a constant and the equations create two parallel lines. Which of the following pairs of equations also creates two parallel lines?

A) $x - 4y = 10$
   $x + 2by = 1$

B) $x - 4y = 10$
   $2x - by = 1$

C) $2x + 4y = 10$
   $2x - by = 30$

D) $2x + 4y = 10$
   $x - by = 30$

Problem 9

$5x + 200 = 75px$

In the given equation, p is a constant. If the equation has no solutions, what is the value of p?

A) $\frac{1}{15}$

B) $-\frac{1}{15}$

C) $\frac{1}{75}$

D) $-\frac{1}{75}$

Problem 10

Function g is defined by $g(x) = (x - 6)(x - 5)(x - 1)$. Function h is defined by $h(x) = g(2x + 10)$. If the graph of $h(x)$ has x-intercepts at $(a, 0)$, $(b, 0)$, and $(c, 0)$, what is the value of $a + b + c$?

Practice Makes Perfect
PROBLEM SET #10

Answers

1. 2
2. 35
3. D
4. A
5. 30
6. C
7. D
8. B
9. A
10. −9

References

Exercise #7

Exercise #5

Exercise #26

Exercise #29

Exercise #11

Exercise #16

Exercise #24

Exercise #32

Exercise #46

Exercise #21

Notes

_____

_____

_____

_____

_____

_____

_____

_____

_____

_____

_____

_____

_____

Glossary

A

Absolute Value: The absolute value of a number is its distance from zero on the number line, regardless of direction. It is always a non-negative number. To enter an absolute value in Desmos, select $|a|$, enter a number or expression, and select $|a|$ again.

Add Item: Use **Add Item** $+$ to add a row or table.

B

Best Fit: A line or curve of best fit most closely matches a set of data points on a graph to show the overall trend.

C

Clearing a Graph: Click ⊠ next to the expression to remove it. To remove all items in the Expression List, select **Edit List** ⚙ and choose **Delete All**.

Color Icon: The **Color Icon** 🖊 appears next to an equation or information in the Expression List. Click it to hide or show the graph. Click and hold to open a panel with options to change the graph's color and style.

Compound Inequalities: Compound inequalities are multiple inequalities combined into one statement, such as $3 < x + 3 < 7$. To solve compound inequalities in Desmos, split them into two separate inequalities, like $3 < x + 3$ and $x + 3 < 7$. Enter each inequality into its own row.

Constant: A constant is a value that does not change. For example, in the linear equation $y = 3x + 5$, 3 and 5 are constants. Sometimes constants are represented by letters, such as a or b. In the linear equation $ax + by = 10$, a, b, and 10 are constants and x and y are variables.

Coordinate Point: A coordinate point is a pair of numerical values that specifies a location on the xy-plane. It is written as (x, y).

D

Default View: Click **Default View** 🏠 at the beginning of each question to reset the Graph Window. If the house icon is missing, the Graph Window is already in the default view.

E

Equation: An equation is a mathematical sentence that shows two expressions have the same value. It usually includes an equal sign ($=$) to state that both sides are equal.

Expression: An expression is a math phrase made up of numbers, variables, or operations, but without an equal sign ($=$).

Expression List: The Expression List is the area on the left side of Desmos where you enter equations and tables.

F

Function: A function is a rule that pairs each input with exactly one value for y. A function can be written using $f(x)$ or $g(x)$ and so on—but in Desmos, it is graphed as y.

Function Table: A Function Table displays a table of values generated from a given function. To create one, enter an equation or function with y isolated (or the function isolated), click **Edit List** ⚙ , then **Create Table** .

G

Graph: In this guide, a graph usually refers to an individual object you create in Desmos, such as a line, parabola, or circle. These appear in the Graph Window. The word graph can also refer to the xy-plane, also called the coordinate plane.

Graph Window: The Graph Window is the xy-plane in Desmos where graphs appear.

H

Hiding and Unhiding Graphs: To hide a graph, click the **Color Icon** next to its row. To make the graph visible again, reclick the **Color Icon**. If two graphs overlap, hide the one on top to reveal the graph underneath.

I

Inequalities: Inequalities are mathematical statements indicating that one quantity is less than or greater than another. For example, $x > 5$ means x is greater than 5, and $2 \leq x < 7$ means x is greater than or equal to 2 and less than 7.

Integer: An integer is a number without fractions or decimals. It can be positive, negative, or zero. Examples include -3, 0, and 3.

Intersection: An intersection is a point where two or more graphs meet.

Isolating y: Isolating y is a strategy to rewrite an equation so the y value is by itself. For example, isolate y in the equation $2x + y = 5$ by rewriting it as $y = 5 - 2x$.

L

Linear Function: A linear function creates a straight line when graphed. It can be written in the form $y = mx + b$, where m is the slope and b is the y-intercept.

M

Maximum Point of a Quadratic: The maximum of a quadratic is the highest y value on its graph, which occurs at the vertex when the parabola opens downward. (For other graphs, the highest point is also called a maximum.)

Mean: The mean, or average, is the sum of a set of numerical values divided by the total number of values.

Median: The median is the middle value of a set of numbers when arranged in ascending or descending order. If the set contains an odd number of values, the median is the middle value. If the data set has an even number of values, the median is the average of the two middle values.

Minimum Point of a Quadratic: The minimum of a quadratic function is the lowest y value on its graph. This occurs at the vertex when the parabola opens upward. (For other graphs, the lowest point is also called a minimum.)

P

Parabola: A parabola is a U-shaped curve that can open upwards or downwards. It is the graph of a quadratic function, which is an equation in the standard form $y = ax^2 + bx + c$, where a, b, and c are constants.

Parallel Lines: Parallel lines have the same slope and never cross. Since they never meet, they do not share any solutions.

Percent: In Desmos, type a percent directly in the Expression List (for example, $x\%$ of $80 = 70$). Desmos will either show the value of x in a gray box or graph it as a vertical line, depending on the question.

Perimeter: The perimeter is the total distance around the edge of a figure.

Points of Interest: Points of interest are coordinate points that Desmos identifies as significant, such as intercepts, intersections, and minimum and maximum points. These points appear as dots on the graph. Clicking these dots reveals their coordinates.

Product: A product is the result of multiplying two or more values.

Pythagorean Theorem: The Pythagorean Theorem states that in a right triangle, the square of the length of the hypotenuse (c) is equal to the sum of the squares of the lengths of the other two sides (a and b). Mathematically, this statement is expressed as $a^2 + b^2 = c^2$.

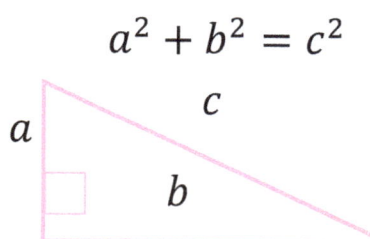

Q

Quadratic Function: A quadratic function in standard form is $y = ax^2 + bx + c$, where a, b, and c are constants and $a \neq 0$. The graph of a quadratic function is a parabola.

R

Real Solutions: Real solutions are values that satisfy an equation and appear on a graph. Imaginary solutions are not covered in this guide. The term **real solution** is used when imaginary ones could exist, but that will not affect how you solve problems in this guide.

Regressions: Regression analysis, or regressions, is a technique used to fit a line or curve to a set of data points to estimate an equation that best represents the trend. Technique #5 uses regressions to analyze data trends and generate equations for lines and parabolas.

S

Scatter Plot: A scatter plot displays coordinates on the xy-plane. These points often do not fit exact lines or curves but instead reveal general trends or patterns in the data.

Slider: The slider tool assigns real, adjustable values to constants, allowing you to test different scenarios and observe changes on the graph.

Slope: Slope is the measure of the steepness of a line, represented by m in the equation $y = mx + b$.

Slope-Intercept Form: The slope-intercept form of a line is written as $y = mx + b$, where m represents the slope of the line and b represents the y-intercept.

Sum: The sum is the result of adding two or more numbers.

System: In this guide, a **system** is a system of equations, unless the context clearly involves inequalities.

System of Equations: A system of equations is a set of two or more equations with the same variables. The solutions are the values that make all of the equations true at the same time. On a graph, the solutions appear where the graphs intersect.

System of Inequalities: A system of inequalities is a set of two or more inequalities with the same variables. The solution is the region where all the inequalities overlap—shown as a double-shaded area on the graph.

T

Tables: There are two types of tables used in this manual. To enter coordinate points, click **Add Item** ⊕ and select **Table** . To generate coordinate points from an equation or function, click **Edit List** ⚙ , then **Create Table** ▦ . (See **Function Table** for more.)

Tilde: The tilde (~) is used to show that the data points are approximations and the equation is a best-fit model. In this guide, it is used in Technique #6.

Truncated Decimals: Desmos shows repeating decimals as truncated, or shortened. If a decimal looks truncated, it's safest to use the fraction form or round your answer to the last place possible (For example, round 4.99999 to 5.) Here are some fractions and how Desmos might show them as decimals.

Fraction	Decimal
$\dfrac{1}{3}$	0.3333
$\dfrac{2}{3}$	0.6666
$\dfrac{1}{6}$	0.1666
$\dfrac{5}{6}$	0.8333
$\dfrac{1}{7}$	0.1428
$\dfrac{2}{7}$	0.2857
$\dfrac{4}{7}$	0.5714
$\dfrac{5}{7}$	0.7142
$\dfrac{1}{9}$	0.1111
$\dfrac{1}{11}$	0.0909

V

Variable: A variable is a symbol, usually a letter, representing a number that can change. Standard variables include x and y.

Vertex: The vertex of a parabola represents either the highest point (maximum) or the lowest point (minimum) on the graph, depending on whether the parabola opens upwards or downwards.

Vertex Form of a Quadratic: Vertex form is $y = a(x - h)^2 + k$, where (h, k) is the vertex of the parabola.

Vertical Lines: Vertical lines on a graph represent solutions where the x-coordinate is constant. The equation for a vertical line is written as $x = c$, where c is a constant. Examples include $x = 7$, $x = 10$, and $x = -2$.

W

Warning Sign: A Warning Sign indicates a potential error. Clicking will reveal information about the error and suggest how to fix it.

X

x-intercept: The x-intercept is the point where a graph crosses the x-axis, represented as $(x, 0)$.

xy-plane: The xy-plane is formed by the x- and y-axes, where points and graphs are plotted. It is also called the coordinate plane. In Desmos, it appears as the Graph Window.

Y

y-intercept: The y-intercept is the point where a graph crosses the y-axis, represented as $(0, y)$.